Activ**Physics 2**
Workbook

Alan Van Heuvelen
The Ohio State University

Paul D'Alessandris
Monroe Community College

 **ADDISON-WESLEY**

An Imprint of Addison Wesley Longman, Inc.

Menlo Park, California • Reading, Massachusetts
New York • Harlow, England • Don Mills, Ontario • Sydney
Mexico City • Madrid • Amsterdam

ACKNOWLEDGMENTS

WORKBOOK

Publisher: Robin Heyden
Sponsoring Editor: Sami Iwata
Project Coordinators: Bridget Biscotti Bradley, Catherine Flack
Accuracy Checkers: Gordon Aubrecht and Mike Ziegler, The Ohio State University
Senior Production Editor: Larry Olsen
Copyeditor: Luana Richards
Composition: London Road Design, Redwood City, CA
Cover Designer: Michael Rogondino
Director of Marketing: Stacy Treco
Channel Marketing Manager: Gay Meixel
Market Development: Andy Fisher

CD-ROM

Publisher: Robin Heyden
Sponsoring Editor: Sami Iwata
Multimedia Production Manager: Erika Bjune
Simulations: OpenTeach Software, Inc.
ActivPad: Sophist Solutions, Inc.
Project Assistant: Kim Askew
Accuracy Checkers: Gordon Aubrecht and Mike Ziegler, The Ohio State University
Designer: Michael Rogondino
Director of Marketing: Stacy Treco
Channel Marketing Manager: Gay Meixel
Market Development: Andy Fisher

Cover Photo: Seyfert Galaxy NGC 4151, Hubble Space Telescope, STIS WFPC2, June 9, 1997
    B. Woodgate (GSFC), J. Hutchings (Dominion Astrophysical Observatory), and NASA.
Unit 17 photo courtesy of Don Eigler, IBM Research Division. Units 13, 16, 18, 19 © 1998 PhotoDisc.

Unit 20 photo courtesy of National Institute of Standards and Technology, from Whitman, L. J.,
    Stroscio, J.A., Dragoset, R.A., and Celotta, R.J., *Phys. Rev. Letts,* 66 (1991):1338.

ISBN 0-201-36111-6 (Workbook with CD-ROM)

ISBN 0-201-35714-3 (Workbook only)

1 2 3 4 5 6 7 8 9 10-CRS-03 02 01 00 99 98

 **ADDISON-WESLEY**

2725 Sand Hill Road
Menlo Park, California 94025

About the Authors

Alan Van Heuvelen, of The Ohio State University, is a respected physics professor, author, and pioneer of active learning methods. His Active Learning Problem Sheets (the ALPS Kits) encourage student participation and learning in large and small classes and while working alone and in small groups. Now these same interactive techniques are easily extended and better visualized with the ActivPhysics simulations and activities.

Paul D'Alessandris has been teaching physics and engineering science at Monroe Community College since 1990. During this time his focus has been on developing curriculum that both incorporates the results of physics education research and effectively deals with a diversity of student abilities. Professor D'Alessandris has received several National Science Foundation grants in support of his work.

11

ELECTRICITY

Questions 1–2 Compare the force magnitudes. Draw arrows representing the force that each charge exerts on the other for the two situations shown. Make the arrows the correct relative lengths.

$\oplus$

$Q_1 = 10 \times 10^{-8}$ C

$\oplus$

$Q_2 = 5.0 \times 10^{-8}$ C

$\oplus$

$Q_1 = 10 \times 10^{-8}$ C

$\ominus$

$Q_2 = -2.0 \times 10^{-8}$ C

Question 3 Test your predictions by adjusting the magnitudes of the charges on the simulation. Then write a general rule that compares the relative magnitude of the force that charge Q_1 exerts on charge Q_2 to the magnitude of the force that Q_2 exerts on Q_1.

Question 4 Dependence of electric force on the signs of the charges. Grab charge Q_2 with the pointer and move it right so that $r_{12} = 200$ cm—five divisions. Leave the value of Q_1 at $+10 \times 10^{-8}$C and vary the value of Q_2. Then leave electric charge Q_2 fixed and vary electric charge Q_1. Indicate both in words and diagrams how the direction of the electric force that one charge exerts on the other depends on the signs of the charges.

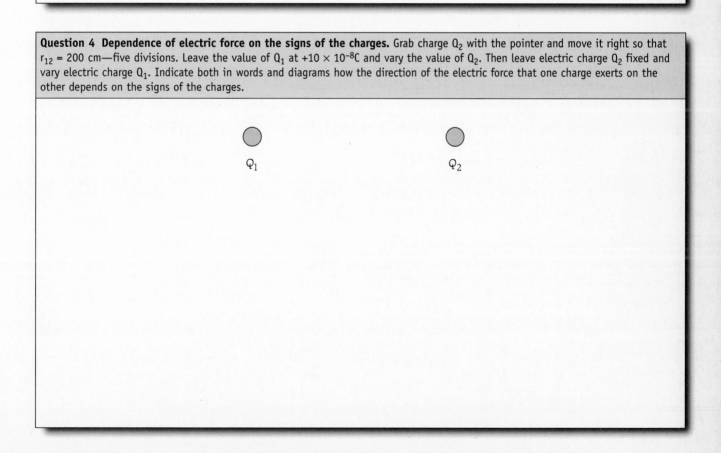

Q_1

Q_2

Question 5 Dependence of electric force on charge separation. Set the values of both Q_1 and Q_2 equal to $+10 \times 10^{-8}$ C. Construct a table that indicates the distance r separating the charges and the magnitude of the electrical force that one exerts on the other. Start with a separation r = 40 cm (0.40 m or one division on the screen) and record the force that one charge exerts on the other. Then try a separation of 80 cm and record the force. Before trying 120 cm, see if you can predict the force. Then try it. Predict the magnitude of the force for separations 160 cm and 200 cm. Finally, devise a rule that indicates how the magnitude of the force that one electric charge exerts on another depends on the distance between the two charges. When finished, compare your thinking to that of the Advisor.

$\oplus$

Q_1

$\ominus$

Q_2

Question 6 Force law between two point charges. Set the values of both Q_1 and Q_2 equal to zero and their separation equal to 100 cm (1.0 m or 2.5 divisions on the screen). Now, increase the value of each charge to the maximum possible and record the magnitude of the electric force that one exerts on the other. Then, leave the separation fixed and try different charge combinations and note the magnitude of the force for each charge combination. Finally, devise a rule that indicates how the magnitude of the force that one electric charge exerts on another depends on the values of the two charges. When finished, compare your thinking to that of the Advisor.

$\bigcirc$

Q_1

$\bigcirc$

Q_2

©1999 Addison Wesley Longman, Alan Van Heuvelen and Paul D'Alessandris

Question 7 Ratio calculation. Change the separation of Q_1 and Q_2 to r_{12} = 200 cm and their charges to Q_1 = +8.0 × 10^{-8} C and Q_2 = −4.0 × 10^{-8} C. Note the electric force that one exerts on the other. Predict the magnitude of the force if you decrease the separation to 100 cm and change the charges so that Q_2 = +2.0 × 10^{-8} C and Q_2 = −8.0 × 10^{-8} C. After your prediction, grab Q_2 and change its separation from Q_2 to 100 cm and adjust the charge-value sliders to the new values in order to check your work.

$\oplus$
Q_1

$\ominus$
Q_2

Question 8 In-your-head ratio calculation. Place charges 1 and 2 in a horizontal line separated by 100 cm. Set Q_1 = +2.0 × 10^{-8} C and Q_2 = +3.0 × 10^{-8} C. Note the magnitude of the electric force that one exerts on the other. Suppose that you increase the separation of the charges to 200 cm. Decide two pairs of values for the charges so that they exert the same sign and magnitude force on each other as before their separation was increased.

$\bigcirc$
Q_1

$\bigcirc$
Q_2

Question 1 Force diagrams. Construct a separate force diagram for each charged ball shown and indicate with an arrow the direction of the net force. Include only the electric forces. The objects are equally spaced.

$\ominus$
$Q_1 = -2q$

$\oplus$
$Q_2 = +q$

$\ominus$
$Q_3 = -q$

$\ominus$
$Q_1 = -2q$

$\oplus$
$Q_2 = +q$

$\ominus$
$Q_3 = -q$

Question 2 Force diagrams. Construct a separate force diagram for each charged ball shown and indicate with an arrow the direction of the net force. Include only the electric forces. The objects are equally spaced.

$\oplus$
$Q_1 = +2q$

$\oplus$
$Q_2 = +q$

$\ominus$
$Q_3 = -q$

$\oplus$
$Q_1 = +2q$

$\oplus$
$Q_2 = +q$

$\ominus$
$Q_3 = -q$

Question 3 With the charges still separated by 200 cm, determine one set of values for the charges so that the net force on $Q_1 = +10 \times 10^{-8}$ C is zero. When finished, adjust the charges in the simulation to check your prediction.

Q_1

Q_2

Q_3

Question 4 Construct a force diagram for each charged ball and estimate the direction of the net electric force on each ball. **Note:** $Q_1 = +2q$, $Q_2 = +2q$, and $Q_3 = -q$. The charges are at the corners of a square 200 cm on each side.

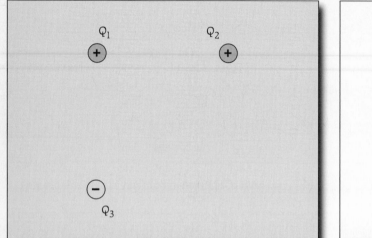

Question 5 Heart dipole. (a) Construct a force diagram for charge Q_3 and determine the direction of the net electric force exerted on Q_3 by the electric dipole if Q_3 has a positive charge and (b) if it has a negative charge. Can you understand how the dipole electric charges on the heart, caused by contracting heart muscles, push oppositely charged ions in the body tissue in opposite directions—like a little battery?

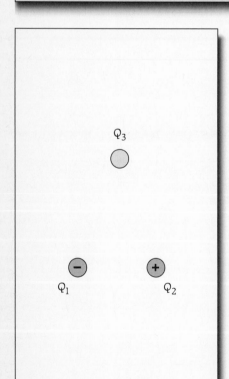

a.

$Q_3 > 0$

b.

$Q_3 < 0$

Question 1 Construct a force diagram for each charged ball shown and determine the net electric force on each ball. The balls are separated by 2.00 m.

$\ominus$	$\oplus$	$\ominus$
$Q_1 = -10.0 \times 10^{-8}$ C	$Q_2 = +5.0 \times 10^{-8}$ C	$Q_3 = -5.0 \times 10^{-8}$ C

$\ominus$	$\oplus$	$\ominus$

Question 2 Construct a force diagram for each charged ball shown and determine the net electric force on each ball. The balls are separated by 2.00 m.

$\oplus$	$\oplus$	$\ominus$
$Q_1 = +10.0 \times 10^{-8}$ C	$Q_2 = +5.0 \times 10^{-8}$ C	$Q_3 = -5.0 \times 10^{-8}$ C

$\oplus$	$\oplus$	$\ominus$

Question 3 Construct a force diagram for each charged ball and determine the net electric force on each ball. **Note:** Q_1 = +10.0 × 10^{-8} C, Q_2 = +10.0 × 10^{-8} C, and Q_3 = −5.0 × 10^{-8} C. The charges are at the corners of a square 1.00 m on each side.

Q_1 Q_2

(+)   (+)

(−)

Q_3

Question 4 Determine the net electric force exerted by the "heart" dipole charge on Q_3 = +3.0 × 10^{-8} C and on Q_3 = −3.0 × 10^{-8} C. The dipole charges Q_1 = +10.0 × 10^{-8} C and Q_2 = −10.0 × 10^{-8} C. The dipole charges are separated by 1.60 m and are 1.44 m from Q_3.

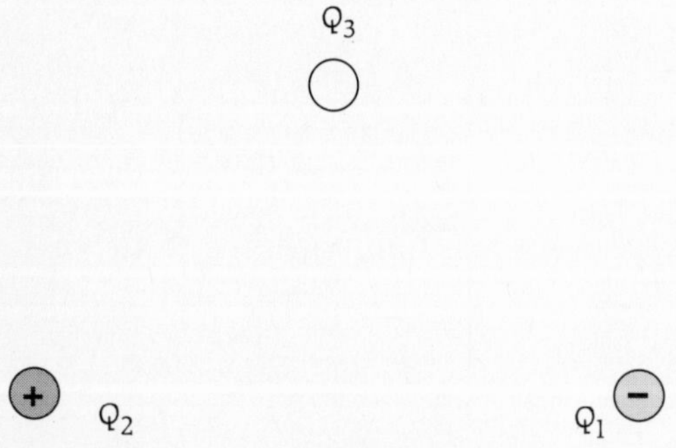

Q_3

( )

(+) Q_2 Q_1 (−)

©1999 Addison Wesley Longman, Alan Van Heuvelen and Paul D'Alessandris

Question 1 Electric field. A positive charge $Q_1 = +10.0 \times 10^{-8}$ C is the source of an electric field. Use a positive test charge $Q_2 = +4.0 \times 10^{-8}$ C to explore the electric field at the positions of the four dots shown in the figure. Summarize your observations in words.

Question 2 Electric field. Determine the direction and calculate the magnitude of the electric field produced by Q_1 at the position 1.0 m to the right of Q_1. On a separate paper, find **E** at the other three positions. Each division on the paper is 0.40 m.

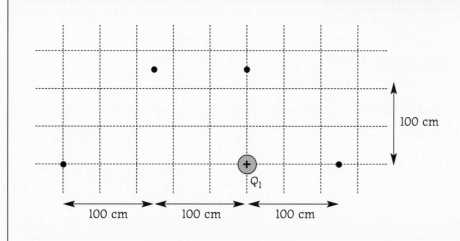

Question 3 Representing the electric field. The lines in the figures below represent the electric field produced by a +2-μC positive charge (left sketch) and a –5-μC charge (right sketch). Develop rules for the way the electric field is represented in these diagrams.

Consider in particular:
- the direction of the lines
- the places where lines start and end relative to the sign of the source charges
- the separation of the lines in a particular region relative to the strength of the field
- the number of lines starting or ending on a charge relative to the magnitude and sign of the charge

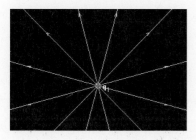 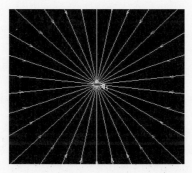

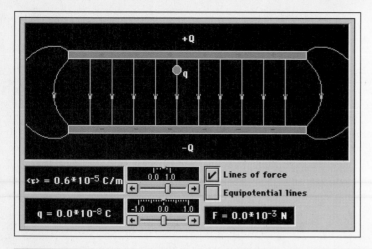

Question 4 The charge q can be placed at different positions between the plates, a region where the field is said to be uniform. Why is the field between the plates called uniform?

Question 5 Is the magnitude of the force exerted by the field on q greater, the same, or less if the charge is moved down near the lower negatively charged plate? Explain.

Question 6 Place the positive charge q in the middle between the plates. What happens to the force on q if you change its sign? Explain. After your prediction, change the sign and magnitude of q and move it to different positions. Note in particular the force on the charge when in the central region where the field is "uniform" compared to the force on the charge in the side fringe regions where the field is not uniform.

©1999 Addison Wesley Longman, Alan Van Heuvelen and Paul D'Alessandris

Question 1 In light pencil, indicate the direction and relative magnitude of the force that each dipole charge exerts on a positive test charge placed at the six positions indicated by the black dots. Graphically, add the forces to estimate the direction and magnitude of the dipole electric field at each position. When finished, move test charge Q_3 in the simulation to see the forces exerted on it. Also, compare your predicted field direction to the field direction shown in Question 2 below.

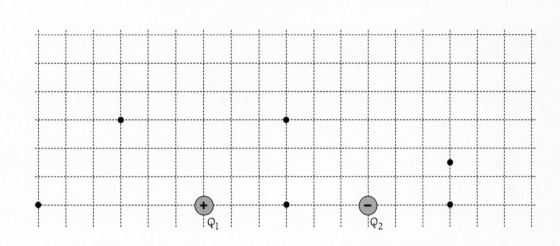

Question 2 The field produced by an electric dipole is shown below. The magnitude of each charge is 3.0 μC and the charged particles are separated by 2.0 m. Indicate (yes or no) if the field line pattern is consistent with the following rules for such representations.

- The electric field direction at a point is in the direction of the field line at that point. _____
- Lines start at positive charges and end at negative charges. _____
- The electric field in a particular region varies continuously and, between lines, points in about the same direction as neighboring lines. _____
- The distance between adjacent field lines at a particular point is proportional to the magnitude of the electric field at that point. _____
- The number of lines leaving or entering a charged particle is proportional to the particle's charge. _____

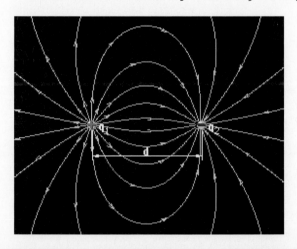

Question 3 Calculate the magnitude and direction of the electric field at the position shown in the diagram.

Your calculation should involve

- a force diagram
- the addition of the force components exerted on a pretend positive test charge +q placed at that point
- the calculation of the net force and the electric field (**E** = **F**/q) at that point

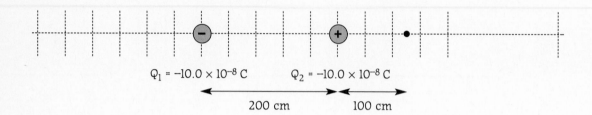

$Q_1 = -10.0 \times 10^{-8}$ C $Q_2 = -10.0 \times 10^{-8}$ C

200 cm   100 cm

Question 4 Calculate the magnitude and direction of the electric field at the position shown in the diagram.

Your calculation should involve

- a force diagram
- the addition of the force components exerted on a pretend positive test charge +q placed at that point
- the calculation of the net force and the electric field (**E** - **F**/q) at that point

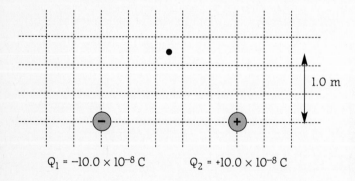

1.0 m

$Q_1 = -10.0 \times 10^{-8}$ C $Q_2 = +10.0 \times 10^{-8}$ C

Question 1 Plotting the electric field. Draw the electric field lines that represent the electric field caused by two equal positive charges.

Question 2 Plotting the electric field. Draw the electric field lines that represent the electric field caused by the two unequal positive charges. The left ball has twice the positive charge as the right ball.

Question 3 If you adjust the charge on ball 1 to +4 µC and the charge on ball 2 to +1 µC, how many field lines emanate from ball 1 compared to ball 2?

Question 4 Where on the x-axis is the electric field zero? The charges are separated by 2.0 m.

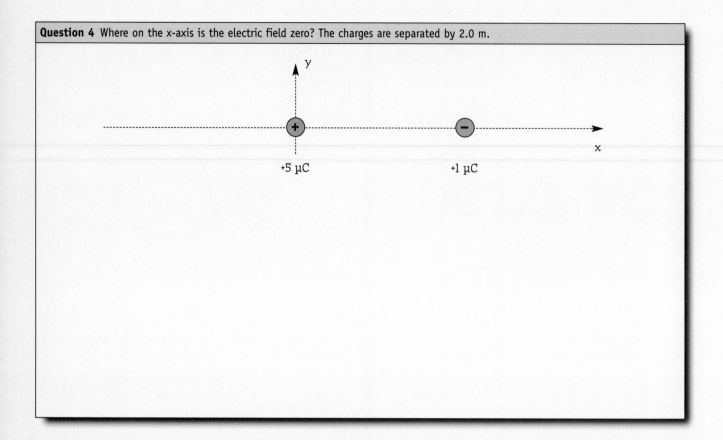

Question 5 Determine the value of Q_2 so that the electric field at the origin is zero. The left charge is 1.50 m from the origin and Q_2 is 1.00 m from the origin.

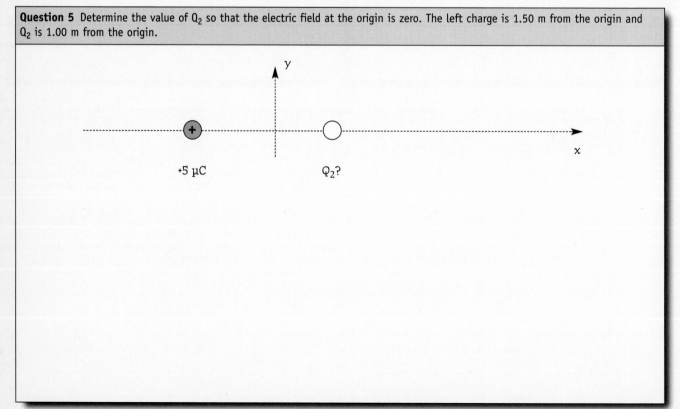

Question 1 Flux into or out of an oval. Design your own experiments to see how electric charge affects the flux into or out of the oval.

In your experiments, you can
• change the shape of the oval
• move the center of the oval so that it surrounds the charge or does not surround the charge
• change the magnitude and sign of the electric charge

When finished, develop in words a qualitative rule to determine the electric flux flowing into or out of the oval. Give examples to support your rule.

Question 2 Electric flux with two charges. In the simulation, click the "two charges" configuration. You can now adjust the sign, magnitude, and separation of two electric charges. Repeat the experiments done in Question 1. Does your rule apply for this two charge system?

You can
• change the shape of the ring
• change the position of the center of the ring (move it all over)
• change the magnitudes and signs of the electric charges

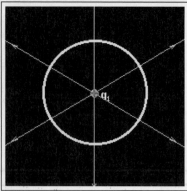

Question 3 First way to determine electric flux. The meter indicates the electric flux ϕ into or out of the oval and the net electric charge q_1 inside the oval. What happens to the flux if you double or triple the positive electric charge inside the oval?

What happens to the flux if you double or triple the negative electric charge inside the oval?

Find an equation with a proportionally constant that relates the electric flux into or out of the oval and the electric charge inside the oval.

Question 4 Second way to determine electric flux. The green electric field lines in the simulation screen represent the electric field surrounding the source charges. Develop a rule for the electric flux passing out of or into the oval by counting the electric field lines passing out of or into the oval.

Question 5 Integrate to find the electric flux leaving a sphere that surrounds a +3.0-µC point charge: $\phi = \iint \mathbf{E} \cdot \mathbf{dA}$. We do this in steps.

(b) For the situation shown in the figure on the left, explain why $\mathbf{E} \cdot \mathbf{dA}$ = E dA for every element on the surface.

(c) Why can you take E outside the integral—the last step below?

$$\text{flux} = \iint \mathbf{E} \cdot \mathbf{dA} = \iint E\, dA = \iint E\, dA$$

(a) What is the direction of **dA**, a small part of the sphere's surface area?

(d) Why can you now simply multiply the magnitude of the electric field at the surface times the total area of the sphere's surface—the last step below?

$$\text{flux} = \iint \mathbf{E} \cdot \mathbf{dA} = \iint E\, dA = \iint E\, dA = E(\text{Area}) = E\,(4\pi r^2)$$

Question 6 Determine the electric flux leaving or entering the oval shown using the charge inside the oval and the number of flux lines crossing the oval surface methods. The charge is +4 μC.

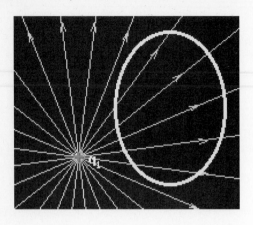

Question 7 Determine the electric flux leaving or entering the oval in the simulation using the charge inside the oval and the number of flux lines crossing the oval surface methods, q_1 = +4 μC and q_2 = −2 μC.

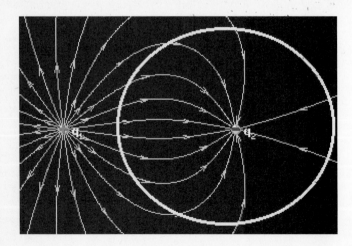

Question 8 Use the charge inside and integral methods to determine the electric flux leaving the 15-mm-radius sphere surrounding a $+10 \times 10^{-10}$ C charge. The electric field at the surface of the 15-mm-radius sphere is 40,000 N/C.

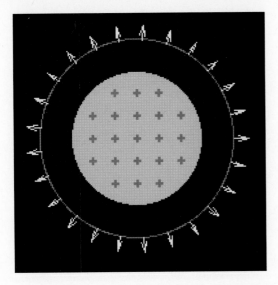

Question 9 Repeat the calculations in Question 8, only this time with the 15-mm-radius sphere surrounding a -10×10^{-10} C charge. The electric field at the surface of the 15-mm-radius sphere is 40,000 N/C and points into the spherical surface.

Question 1 Electric field outside a charged sphere. A 10-mm-radius solid sphere has $+10 \times 10^{-10}$ C of electric charge distributed uniformly throughout the sphere. Use Gauss's law to determine the electric field caused by this charge at a distance of 15 mm from the center of the sphere. $\iint \mathbf{E} \cdot \mathbf{dA} = 4\pi\, Q_{inside}$. We do this in steps.

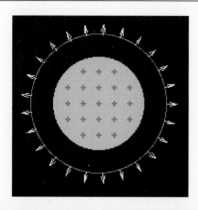

(b) Determine the charge inside the Gaussian surface Q_{inside}.

(a) First, determine an expression for the electric flux out of the Gaussian surface.

- What is the direction of **dA**, a small part of the Gaussian sphere's surface area?

- For the situation shown above, *explain why* **E** • **dA** = E dA for *every* element on the surface.

- Why can you take E outside the integral—the last step below?

flux = $\iint \mathbf{E} \cdot \mathbf{dA} = \iint E\, dA = \int E\, dA$

- Why can you now simply multiply the magnitude of the electric field at the surface times the total area of the sphere's surface—the last step below?

flux = $\iint \mathbf{E} \cdot \mathbf{dA} = \iint E\, dA = \int E\, dA = E(\text{Area}) = E(4\pi R^2)$

(c) Insert the electric flux and Q_{inside} into Gauss's law and find the electric field.

Question 2 Electric field outside a charged spherical shell. A 10-mm-radius spherical shell, like a basketball, has $+10 \times 10^{-10}$ C of electric charge distributed uniformly on its surface. Use Gauss's law to determine the electric field caused by this charge at a distance of 15 mm from the center of the sphere. $\iint \mathbf{E} \cdot \mathbf{dA} = 4\pi k\, Q_{inside}$

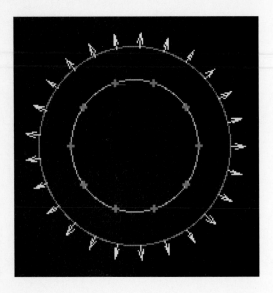

Question 3 Electric field inside the charged spherical shell. For the charged spherical shell in Question 2, use Gauss's law to predict the value of the electric field inside the shell. If you want, you can calculate the magnitude of the electric field at R = 5 mm from the center of the 10-mm-radius spherical shell. $\iint \mathbf{E} \cdot \mathbf{dA} = 4\pi k\, Q_{inside}$

Question 4 Electric field inside a charged solid sphere. Predict the value of the electric field 5.0 mm from the center of a 10-mm-radius solid sphere. The sphere is uniformly charged with $+10 \times 10^{-10}$ C of electric charge.

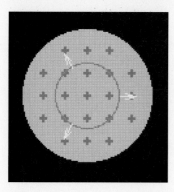

(a) First, determine the electric charge inside (Q_{inside}) a 5.0-mm-radius Gaussian surface. To do this, find the charge density in the bigger 10-mm-radius sphere, the volume of the 5-mm radius Gaussian sphere, and then the charge inside that Gaussian sphere.

(b) Then use Gauss's law to determine the electric field 5.0 mm from the center of the charged sphere. $\iint \mathbf{E} \cdot \mathbf{dA} = 4\pi k\, Q_{inside}$

Question 5 Electric field inside a charged solid sphere. Show that the electric field INSIDE the solid uniformly charged sphere varies as

$$E = \left(\frac{kQ}{R_{\text{charged sphere}}} \right)^2 \left(\frac{R}{R_{\text{charged sphere}}} \right)$$

where $R_{\text{charged sphere}}$ is the radius of the charged sphere (10 mm in the simulation), Q is the total charge on the sphere (adjustable with the Q slider), and R is the distance from the center of the sphere to the point where the electric field is being calculated.

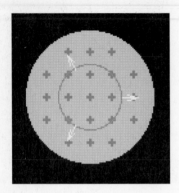

Question 1 Electron path. A negatively charged electron moves through a uniform electric field that points up. The electron enters the field moving in the horizontal direction; draw a line showing its subsequent motion after entering the field.

Question 2 x Velocity component. How does the x-component of velocity change as the electron moves through the vertical electric field?

Question 3 Electron path. Make the electron's x-component of initial velocity positive and its y-component of initial velocity zero.

- Construct a force diagram for the electron when passing through the electric field.
- Apply the y-component form of Newton's second law for the electron.
- Calculate the electron's acceleration in the y-direction.
- Run the simulation. Use the simulation y-component and time numbers to independently calculate the electron's acceleration in the y-direction. Compare these numbers to your previous Newton's second law prediction.

Question 4 Time of flight. Note the time interval needed for the electron to hit the lower plate.

If you reduce the initial x-component of velocity by one-half, is the time of flight: **(a)** less? **(b)** the same? **(c)** more? Explain.

Question 5 Effect of electric field on motion. If you reduce the electric field by half, what is the effect on

(a) the electron's acceleration in the x-direction?

(b) the electron's acceleration in the y-direction?

(c) the time of flight (the time to reach the bottom plate)?

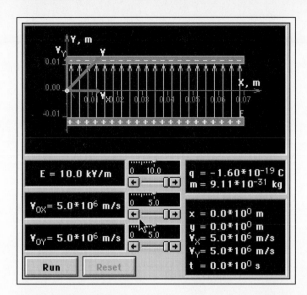

Question 1 Maximum y-position. Predict the maximum y-position of the electron. Does the initial horizontal velocity component affect the outcome?

Question 2 Time of flight. With the same settings as in Question 1, determine the time of flight for the electron (the time interval to reach the bottom plate).

Question 3 Maximum time of flight. Adjust the electric field and the initial y-velocity component so that the electron takes the longest time to return to y = 0.0 m. Note that its flight stops if it hits the upper plate at y = 0.010 m.

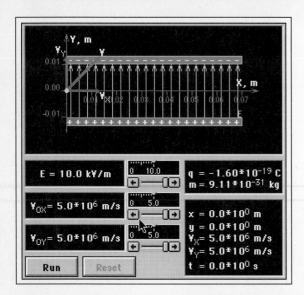

Question 4 Range. With E = +1.0 kV/m and v_{ox} = 5.0 x 10^6 m/s, predict the initial vertical velocity v_{oy} component that causes the electron on the first try to reach position x = 0.07 m and y = 0.00 m.

Question 5 Another problem. (a) Set the electric field to + 1.0 kV/m and the initial horizontal velocity component to +5.0 × 10^6 m/s. Determine the initial vertical velocity component so that the electron reaches x = 0.07 m and y = −0.01 m (on the first try). **(b)** If you double the value of the answer to (a), what value of the horizontal velocity component is needed to reach that same point? Do this work on a separate paper.

Question 6 A final problem. Set the electric field to + 10.0 kV/m and the horizontal and vertical velocity components to +5.0 × 10^6 m/s. Where will the electron hit the plate?

©1999 Addison Wesley Longman, Alan Van Heuvelen and Paul D'Alessandris

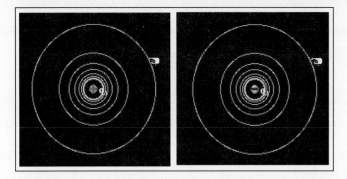

Summarize your observations for both Questions 1 and 2 in the space below.

Question 1 Electric potential due to positive charge. A +3.0-µC point charge produces the circular equipotential lines shown in the left figure above. The electric potential (sometimes called voltage) along each line is the same. The lines are analogous to constant elevation lines on a topological map.

• Move the pointer to one line and press down on the mouse to see the electric potential on that line. Continue moving along the line with the finger down and note the potential at other points on the line. What would an analogous hike in the mountains be like?

• Now, place the pointer near the positive charge and move away from it. What would an analogous hike in the mountains be like?

Question 2 Electric potential due to negative charge. Change the electric charge to −3.0 µC and repeat the two activities in Question 1 for a negative electric charge. In particular, how would you describe the quantity electric potential to another student? Indicate any analogies between charge and electric potential and large massive objects such as the earth and gravitational potential (gy).

Question 3 Electric potential due to an electric dipole. A +3.0-μC charge on the left and a −3.0-μC charge on the right produce the equipotential lines shown. The finger is on a +20-mV line.

- Move the pointer along several lines and note the potential at different positions along a line. What would an analogous hike in the mountains be like?
- Now, place the pointer near the positive charge and move it directly toward the negative charge. What would an analogous hike in the mountains be like?
- What is the potential half way between the charges?

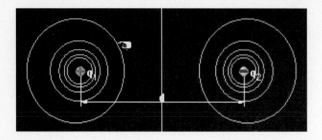

Question 4 Determining the electric potential. Suppose you knew the potential at every point in space that is caused by the positive charge if there alone and by the negative charge if there alone. How would you determine the potential when both charges are present? Note that the potential at the point where the finger tip is located in the sketch in Question 3 is +20 mV. If the positive charge was there alone, the potential would be +27 mV at that point. What do you think the potential would be at that point if the negative charge was there alone?

Question 5 Electric potential due to two positive charges. Suppose you have two +3.0 μC in a region, as shown. Choose several special points on the simulation window and measure the potential at those points with both charges present. Then reduce the right charge q_2 to zero and measure the potential at these points again. (**Warning:** The positions of the equipotential lines change, so they cannot be used to identify special points on the simulation window.) Now predict the potential at these points if the left q_1 charge is zero and the right charge q_2 is returned to +3 μC. Check your predictions by changing the charges and measuring the potential with the finger pointer.)

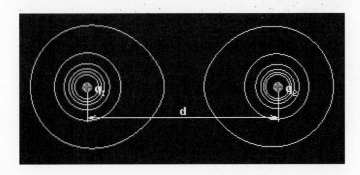

Question 6 Potential due to multiple charges. In the sketch below, you see equipotential lines caused by positive charges +Q distributed uniformly on the top metal plate and by negative charges –Q distributed uniformly on the lower metal plate. In the region between the plates, the equipotential lines are evenly spaced. The lines near the top positively charged plate are at higher potential than the lines near the bottom negatively charged plate. The line in the middle is at zero potential.

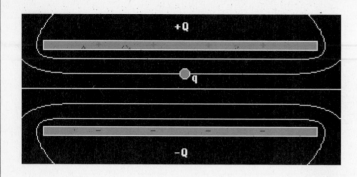

What is the direction of the force on a positively charged (q) ball when it is near the upper plate?

Is the force on that positively charged ball greater, the same, or less if it is closer to the lower plate (at lower potential)? Explain.

Does the magnitude of the electric force exerted on the charge q depend on the value of the potential or on some other property of the potential? Describe the property.

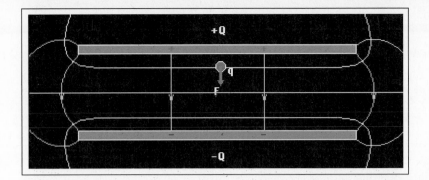

Question 1 Parallel plate electric field lines, equipotential lines, and electric force. The figure above shows a charge density of $+0.2 \times 10^{-8}$ C/m^2 distributed uniformly on the top plate and a charge of -0.2×10^{-8} C/m^2 distributed uniformly on the lower plate. The vertical electric field lines and horizontal equipotential field lines between the plates are also shown, as is the electric force exerted on a positive charge between the plates. On a separate paper, draw the **E** lines, the equipotential lines, and the force on the same charge if the magnitude of charge density is changed to 0.4×10^{-8} C/m^2 and then again to 0.6×10^{-8} C/m^2.

Question 2 Point charge electric field lines and equipotential lines. Are the representations of the electric field lines and the equipotential lines consistent with the following rules?

- Equipotential lines are perpendicular to electric field lines. _____

- The electric field at a position depends on how the potential is changing rather than on the value of the potential at a particular position. _____

- The electric field is greater in the regions where the equipotential lines are closer to each other. _____

- The electric field points "downhill," that is, from higher to lower electric potential. _____

Question 3 Electric force and relation between field and potential. Devise and describe some experiments to decide whether the simulation is consistent with the following relations: $F_y = qE_y$ and $E_y = -\Delta V/\Delta y$.

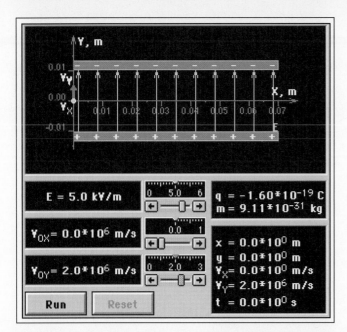

Question 1 Potential difference and electric field. Use the equation $E_y = -\Delta V/\Delta y$ to determine the potential difference between the lower plate and the upper plate.

Question 2 Changes in electric potential and electric potential energy. Suppose the electron moved from its present starting position between the plates to the top plate. Determine its change in electrical potential energy $U_q = qV$.

Question 3 An energy transformation problem. Our goal is to predict the initial vertical speed the electron needs so that it just reaches the top plate but does not hit it. Construct a qualitative energy bar chart for a process in which the electron starts with upward velocity and stops at the instant it reaches the top plate. Assume that the electrical potential energy is zero when the electron is at y = 0.00 m. The possible types of energy are kinetic (K), gravitational potential energy (U_g), elastic potential energy (U_s), and electrical potential energy (U_q).

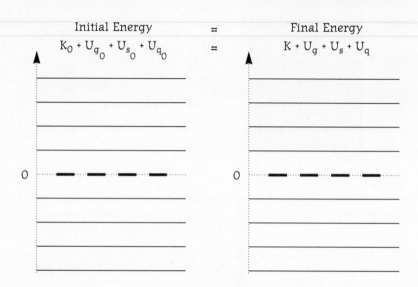

Speed to reach top plate. Use an energy approach and the energy bar chart to determine the initial y-component of velocity that is needed for the electron to just reach the top plate (its speed should be zero when y = 0.010 m). Use the conditions from Questions 1–3.

Question 4 Another problem. Start the electron moving upward at 5.0×10^6 m/s. Use an energy approach to determine the potential difference from its starting position to the top plate so that the electron stops at the top. Then determine the electric field that would produce this potential difference in a distance of 0.010 m.

Question 5 A final problem. Set the initial vertical velocity component to $+3.0 \times 10^6$ m/s and the initial horizontal velocity component to $+5.0 \times 10^6$ m/s. Use an energy approach to determine the electric field that will just deflect the electron when it reaches the top plate.

dc CIRCUITS

12

Hints for Using the Electric Circuit Construction Tools

◆ To add one of the elements to the circuit, click on it in the toolbox. Release the mouse and move the mouse between two dots in the circuit-building area. Then click and the element should appear between the dots.

◆ To remove an element, click on it.

The simulation has tools on the right side that can be used to construct an electric circuit. The tools, starting at the top, are

The Wire Tool makes a wire.

The Switch Tool produces an open switch. The switch can be closed and opened by clicking on it only while the simulation is running.

The Bulb Tool creates a 1.0-ohm resistance bulb.

The Capacitor Tool has two equally long parallel vertical lines and produces the capacitors used in Activities 12.6 and 12.7.

The Battery Tool is a source of potential difference. You can adjust the potential difference (often called voltage) across the battery terminals with a slider.

The Ammeter and Voltmeter Tools create meters that measure the current along a circuit and the potential difference (voltage) across a circuit element, respectively.

The Grounding Tool has an elbow and can set the electric potential in the circuit to zero at some point of your choice.

The Pointer Tool (an arrow) selects some element to adjust its value. The pointer can also indicate the voltage at different points in a circuit if the circuit is grounded with the Grounding Tool. Just click the pointer down at points in the circuit and the voltage is shown in the bottom-right panel of the simulation screen.

Question 1 First circuit. Consider the circuit as shown below. Adjust the battery voltage across the bulb to 10.0 V and observe the brightness of the bulb. How does the bulb brightness change as the voltage across the bulb is decreased?

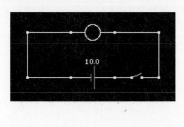

Question 2 Bulb resistance. Adjust the battery voltage and observe the brightness of the bulb and the electric current through the bulb. Complete the following table showing the voltage V across the bulb (for this circuit, the same as the voltage across the battery) and the electric current I passing through the bulb. Use this data to determine the electric resistance R in ohms of the bulb: R = V/I .

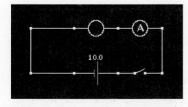

Voltage (V)	10	8	6	4	2	0
Current (A)						

Question 3 Two bulbs in series. Consider the circuit shown below. Which bulb will be brightest when the switch is closed? Add more bulbs in series, and compare the brightnesses of each bulb with others in the circuit. What do your observations tell you?

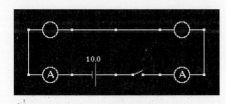

Question 4 Equivalent resistance for series resistors. With the circuit shown below, keep the potential difference across the battery at 10 V. In the table, record the current through the circuit as you replace wires with more 1.0-Ω bulbs in series. Based on your observations, state a rule for the equivalent resistance of a group of series resistive elements (like the bulbs).

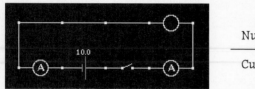

Number of bulbs	1	2	3	4
Current (A)				

The **equivalent resistance** of the bulbs is $R_{equivalent} = V / I$ where V is the potential difference across all of the resistive elements (the bulbs in this case) and I is the current through them. A single resistor with this equivalent resistance would have the same current passing through it if the same potential difference was placed across it.

Question 5 More series resistors. Use the idea for the equivalent resistance of series resistors to predict the current through the circuit shown below for different potential differences across the resistors. The resistors in this case have different resistances. You will have to adjust the resistances in the simulation.

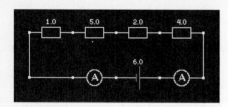

Voltage (V)	10	5	0
Current (A)			

Question 1 Parallel bulbs. With one switch closed, note the brightness of the bulb. How will the brightness of that bulb change if the switch for the second bulb is closed? Compare their brightnesses. How will the brightnesses of the two lower bulbs change if the switch for the third bulb is closed? How do the brightnesses of the three bulbs compare?

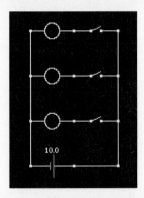

Question 2 Current through different branches.

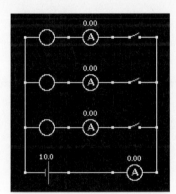

Predict the current through all ammeters when the switch for the lowest bulb is closed. After your prediction, close the switch to see how you did.

Predict the current through all the ammeters when the switches for the two lowest bulbs are closed. After your prediction, close the switches to see how you did.

Finally, predict the current through all the ammeters when the switches for all three bulbs are closed. After your prediction, close the switches to see how you did.

Question 3 Junction rule.

Turn on the switch for the lowest bulb. Look at the two middle junctions along the right side of the circuit (middle two dots on the right side of the circuit), How does the current flowing into each junction compare to the current flowing out of each junction?

Turn on the switches for the two lowest bulbs. How does the current flowing into each of the middle two junctions on the right side of the circuit compare to the current flowing out of each of these junctions?

Turn on the switches for all three bulbs. How does the current flowing into each of the middle two junctions on the right side of the circuit compare to the current flowing out of each of these junctions?

Question 3 Junction rule continued.

Write a general rule that seems to be consistent with your observations for the two junctions just analyzed.

Question 4 Equivalent resistance of parallel resistors. Complete the following table, which shows the current through the 5.0-V battery when placed in parallel across one or more 1.0-Ω bulbs. Use these numbers to write an equation for the equivalent resistance of the parallel resistors. In general equivalent resistance is defined as $R_{equivalent} = V / I$ where V is the voltage across the resistors and I is the total current flowing through them.

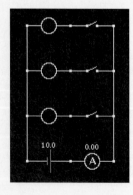

Bulbs in circuit	1	2	3
Voltage (V)	10	10	10
Current (A)			
$R_{equivalent}$ (Ω)			

Question 5 The battery—a constant current source or a constant voltage source? Based on your observations in this and the previous activity, would you say that the battery is a constant potential difference (constant voltage) source or a constant current source? Provide an example to support your choice.

Question 6 Using the equivalent resistance and junction rules. Apply the equivalent resistance rule you developed in Question 4 to predict the total current through the circuit shown below.

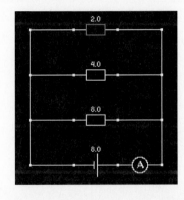

Question 7 Junction rule. Is the junction rule consistent with the current readings for the circuit shown here? Provide two examples.

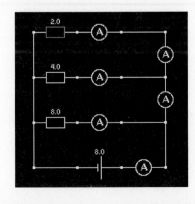

Puzzle 1 Rate the bulbs in this circuit according to brightness, listing the brightest bulb first. Indicate whether any bulbs are equally bright.

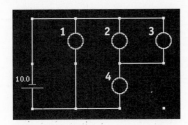

——— ——— ——— ———
Brightest   Dimmest

Puzzle 2 Rate the bulbs in this circuit according to brightness, listing the brightest bulb first. Indicate whether any bulbs are equally bright.

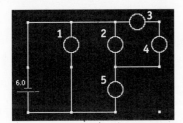

——— ——— ——— ——— ———
Brightest   Dimmest

Puzzle 3 Rate the ammeters in this circuit according to current flowing through them, the largest current ammeter first. The ammeters have zero resistance. Indicate whether any currents are equal. (**Hint.** It might be easier to visualize the circuit if you redraw it without the ammeters.)

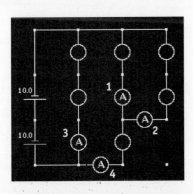

——— ——— ——— ———
Largest   Smallest

Puzzle 4 When the switch is open, three bulbs shine with equal brightness. Indicate how the brightness of each bulb changes when the switch is closed. Justify your choices.

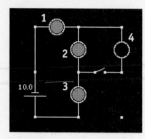

Bulb 1: _____ Becomes brighter
_____ Remains same
_____ Becomes dimmer

Bulb 2: _____ Becomes brighter
_____ Remains same
_____ Becomes dimmer

Bulb 3: _____ Becomes brighter
_____ Remains same
_____ Becomes dimmer

Puzzle 5 Indicate how the brightness of each bulb changes when the switch is closed. Justify your choices.

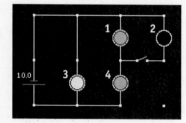

Bulb 1: _____ Becomes brighter
_____ Remains same
_____ Becomes dimmer
Explanation:

Bulb 2: _____ Becomes brighter
_____ Remains same
_____ Becomes dimmer
Explanation:

Bulb 3: _____ Becomes brighter
_____ Remains same
_____ Becomes dimmer
Explanation:

Bulb 4: _____ Becomes brighter
_____ Remains same
_____ Becomes dimmer
Explanation:

Puzzle 6 Three separate circuits each have one bulb. Rate the circuits according to the brightness of the bulb in the circuit, listing the brightest bulb circuit first. Indicate whether any bulbs are equally bright.

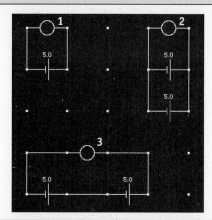

_____ _____ _____

Brightest   Dimmest

Puzzle 7 (a) Determine the current in each ammeter and the relative brightness of the four 1-Ω bulbs when the switch is open. Be sure to note that the left battery has a potential difference of −10 V.

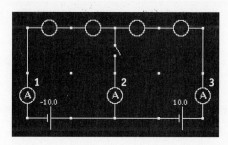

(b) Determine the current in each ammeter and the relative brightness of the four 2-Ω bulbs when the switch is closed. Note that the left battery has a potential difference of −10 V.

Puzzle 8 **(a)** Rate the relative brightnesses of the 1-Ω bulbs.

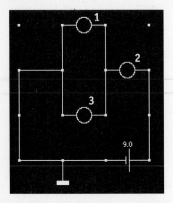

(b) Determine the voltage at each point in the circuit. The circuit is grounded (zero volts) on the left side of the battery.

Measuring Electric Current with an Ammeter

- An ammeter measures the flow of electric charge.

- To measure the current in a branch of a circuit, it must be inserted in series into the circuit branch so that all of the current passing through that branch must also pass through the ammeter.

- If inserted in the branch of a circuit through which we want to measure the electric current, the ammeter must not affect that current flow. Consequently, the ammeter has very low electric resistance (ideally, zero resistance).

- Both the correct and incorrect ways to insert the ammeter in the circuit are shown here. If used as shown on the right, a very large current will flow through the low resistance ammeter and possibly damage it (or blow its fuse).

Correct Ammeter

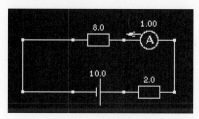

Incorrect Ammeter

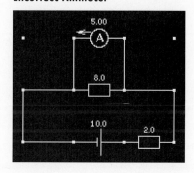

Measuring Potential Difference with a Voltmeter

- A voltmeter measures the potential difference between two points in an electric circuit.

- To measure the potential difference, the voltmeter terminals are touched to the two points. The voltmeter is in parallel with the circuit elements across which the potential difference is being measured.

- If placed in parallel across some part of the circuit, we do not want the voltmeter to provide an alternative path for the circuit's electric current. Consequently, the voltmeter should have somewhat higher electric resistance than the circuit elements across which the terminals are placed—ideally, the voltmeter has infinite resistance.

- Both the correct and incorrect ways to use a voltmeter to measure potential difference are shown here.

Correct Ammeter

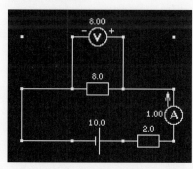

Incorrect Ammeter

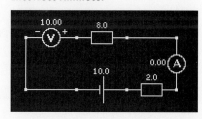

SUMMARY: Kirchhoff's Laws

Loop rule. The net change in electric potential (the voltage change) around any closed loop is zero. This is because the electric potential at every point in the circut can have only one value. A charged particle returns to the same electric potential if it returns to the same point in a circuit after a trip around a loop.

Junction rule. The electric current entering any point in a direct-current (dc) electric circuit equals the electric current leaving that point. If this were not true, electric charge would accumulate or dissipate at that point. This happens on capacitor plates but not at dc circuit junctions.

Question 1 The loop rule. (a) Use the loop rule to predict the electric current through the circuit shown below. **(b)** Then, determine the electric potential (voltage) at each point in the circuit..

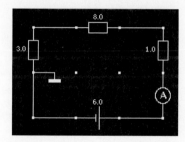

Question 2 (a) Apply the loop rule twice and the junction rule once to get three equations that can be used to determine the ammeter current readings for the circuit shown below. **(b)** Then, determine the electric potential at each point in the circuit.

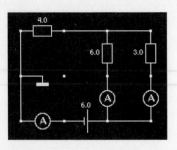

Question 3 (a) Apply the loop rule twice and the junction rule once. Be sure to note that the top battery produces −6.0 V (the left side is the positive terminal). **(b)** Use the three independent equations developed in part (a) to determine the current in each branch of the circuit.

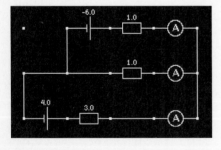

©1999 Addison Wesley Longman, Alan Van Heuvelen and Paul D'Alessandris

Question 4 The circulatory system. Here, we use a very simple electric circuit as an analog for the heart (the battery) and the vessels in the circulatory system (the resistors). The electron current is the "blood flow."

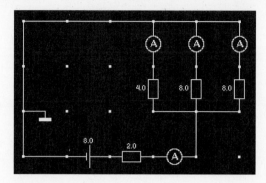

(a) Predict the equivalent resistance of this circuit. Then calculate the battery current and the current through each resistor. Also, determine the electric potential at every position in the circuit.

(b) Suppose the "person's" aorta (the large vessel that leaves the heart—it has a 2-Ω resistance in our simulation) becomes clogged, causing its resistance to blood flow to quadruple (to 8.0 Ω in the simulation). Now, determine the current flow (blood flow) in the other parts of the circuit.

Question 5 A circuit with three identical bulbs is shown below. Run the simulation with the switch open. Then answer the following questions before closing the switch.

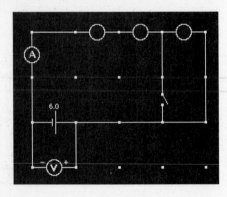

After you have answered the questions, run the simulation and turn the switch on and off to see how you did.

(a) What happens to the brightness of each bulb when the switch is closed?

(b) What happens to the current through the battery when the switch is closed?

(c) What happens to the potential difference across the battery?

(d) What happens to the total power use of the bulbs when the switch is closed?

Question 6 Blowing a fuse. The circuit shown here is analogous to a line in your home that is connected to a fuse that blows if the current exceeds 25.0 A.

As you prepare dinner, you have the following items connected across a 10-V potential difference:

10-Ω light

1-Ω burner

5-Ω mixer

1-Ω oven

5-Ω crock

4-Ω light

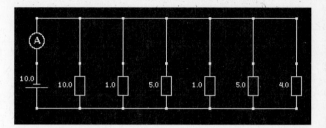

(a) Will you blow the fuse?

(b) If the current does exceed 25 A, remove one or more of the appliances so that the current is 25 A or less but as close to 25 A as possible. Which appliances will you remove?

Capacitance C A capacitor consists of two conducting surfaces separated by a nonconducting region. The capacitance C of the capacitor is a measure of its ability to store opposite electric charge (+q and −q) on the conducting surfaces when a potential difference V is placed across the surfaces. In particular, C = q/V.

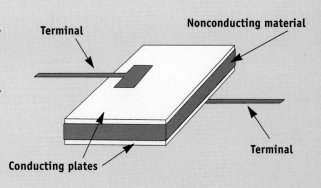

Terminal

Nonconducting material

Terminal

Conducting plates

Question 1 Capacitor in an electric circuit. Predict how the brightness of each bulb, the current reading in each ammeter, and the voltage across the capacitor change with time as you run the simulation. Remember that the capacitor has two very large metal plates that do not touch. Thus there is a gap in the circuit—the space between the plates. After your predictions, run the simulation.

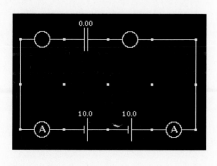

Question 2 Review questions.

(a) Does an open gap between the conducting plates of a capacitor mean that current cannot flow in other parts of the circuit or in that branch of the circuit?

(b) What happens on the capacitor plates when this current flows?

(c) Why does the current flow decrease over time in this particular situation (but not in general)?

Question 3 A capacitor in parallel with a bulb. With the switch open in the circuit below, note in the simulation the ammeter readings and bulb brightnesses for each bulb. If you were to close the switch, there would now be a capacitor in parallel with the bulb on the right (don't close the switch yet).

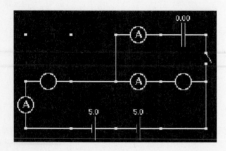

Immediately after closing the switch, would the brightness of the bulb on the right in parallel with the capacitor increse, decrease, or remain the same? Explain.

Immediately after closing the switch, would the brightness of the bulb on the left increase, decrease, or remain the same? Explain.

A long time after closing the switch, predict the relative brightnesses of the bulbs. Explain your reasoning.

After your preditions, close the switch and see what happens. You may have to open the switch, reset the simulation and try the experiment again while focusing on one bulb at a time.

Question 4 Bulb between two capacitors. This circuit has three bulbs and two capacitors all in series across a 30-V potential difference (three 10-V batteries in series). Predict what happens to each bulb immediately after the simulation begins to run. Justify your predictions.

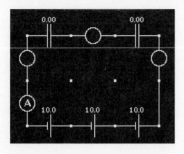

Predict what happens to each bulb 10 or 20 s after the simulation starts. Justify your predictions, then run the simulation to check your reasoning.

Question 5 Three bulbs and a capacitor. Predict the relative brightnesses of the three bulbs and the approximate voltage across the capacitor shown below immediately after the simulation starts to run. Justify your predictions.

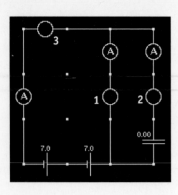

Repeat your predictions, only this time about 3 s after the simulation starts to run. Justify your predictions.

Repeat your predictions, only this time after the simulation has run 10 or 20 s. Justify your predictions.

Question 1 Series capacitors. Run the simulation and decide whether the group of three capacitors in series has more, the same, or less capacitance than the circuit with a single capacitor. All capacitors have the same capacitance, the bulbs the same resistance, and the ammeters zero resistance. Justify your conclusion.

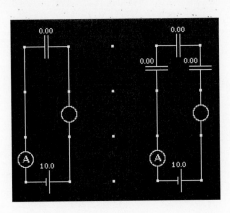

Question 2 Parallel capacitors. Run the simulation and decide whether the group of three capacitors in parallel has more, the same, or less capacitance than the circuit with a single capacitor. All capacitors have the same capacitance, the bulbs the same resistance, and the ammeters zero resistance. Justify your conclusion.

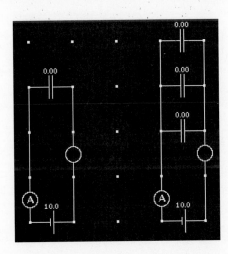

Question 3 Answer the following questions about the circuit shown below.

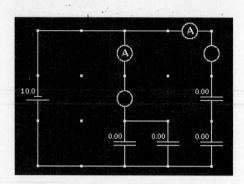

(a) Compare the relative brightness of the two bulbs shortly after the switch is closed. Explain your reasoning.

(b) Compare the relatve brightness of the two bulbs after the capacitors are completely charged. Explain why the brightnesses are as observed.

(c) Determine the potential difference across each capacitor after it becomes fully charged.

Question 4 Answer the following questions about the circuit shown below.

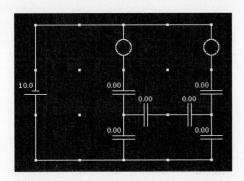

(a) Compare the relative brightness of the two bulbs shortly after the switch is closed.

(b) Compare the relative brightness of the two bulbs after the capacitors are completely charged.

(c) Determine the potential difference across each capacitor after it becomes fully charged.

Question 1 Discharging a capacitor. Set the initial charge on the capacitor to 4.0×10^{-4} C, the resistor resistance to 1.5 MΩ, and the capacitor's capacitance to 9.6 µF. Run the simulation and record the charge on the capacitor each 10 s starting at time zero. State in the form of a rule any systematic variation that you observe in the charge on the capacitor.

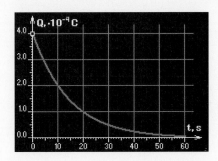

Question 2 Effect of R and C on half-life. With the resistor resistance at 1.5 MΩ and the capacitor's capacitance at 9.6 μF, the half-life for the discharge of the capacitor is 10 s.

(a) Increase the resistance to 3.0 MΩ. Find a value for the capacitance so that the half-life is again 10 s.

(b) Adjust the capacitance again to get a 10-s half-life only this time with the resistor resistance at 6.0 MΩ.

(c) Return the resistance to 3.0 MΩ and the capacitance to 4.8 μF. Increase the resistance and observe the graph. Does the half-life increase or decrease as the resistance increases? If you double the resistance, how does the half-life change?

(d) Return the resistance to 3.0 MΩ and the capacitance to 4.8 μF. Increase the capacitance and observe the graph. Does the half-life increase or decrease when the capacitance increases? If you double the capacitance, how does the half-life change?

(e) Now, write an equation that indicates how the half-life for capacitor discharge depends on the values of the resistance and the capacitance in the RC circuit.

(f) Does the half-life depend on the initial charge on the capacitor?

Question 3 Charge and current. Start with an initial charge of 4.0×10^{-4} C, a resistance of 3.0 MΩ, and a capacitance of 4.8 µF. The half-life is 10 s and the equation for the charge on the capacitor as a function of time is $Q = Q_0 \, e^{-0.693 \, t \, / \, R \, C}$. Determine the current ($i = dQ/dt$) through the resistor as a function of time. Then, determine the value of the current at times 0, 10, 20, and 30 s. Run the simulation to check your work.

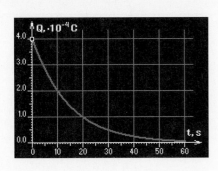

Question 4 Loop rule during discharge. Start with an initial charge of 4.0×10^{-4} C, a resistance of 3.0 MΩ, and a capacitance of 4.8 µF. In the space below, draw a series RC circuit. Assume that the capacitor is initially charged. Use the numbers on the simulation screen to see whether Kirchoff's loop rule applies to the discharge of this RC circuit at times 10, 20, and 30 s after the discharge starts.

13

MAGNETISM

Question 1 Direction of the magnetic field. What will the magnetic field look like when positive current flows through the wire? (Positive current is defined to flow out of the screen.) Sketch the field.

⊙

Question 2 Orientation of the magnetic field. What angle does the magnetic field make relative to the position vector connecting the wire to the point of interest? Answer below, and indicate this angle on your sketch in Question 1.

Question 3 Magnitude along a radial line. Does the magnitude of the field change along a line extending radially away from the wire? If so, describe the change.

Question 4 Magnitude along a field line. Does the magnitude of the field change along the circular field lines? If so, describe the change.

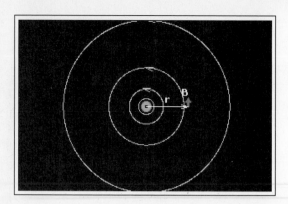

Question 5 Dependence on current. What happens to the magnitude and direction of the magnetic field, at the point in space you are examining, if the current is increased?

Question 6 Flipping the current. What happens to the magnitude and direction of the magnetic field, at the point in space you are examining, if the current is flipped from positive to negative?

Question 7 Iron fillings pattern.

Sketch the pattern formed by the iron filings for a negative current.

Sketch the pattern formed by the iron filings for a positive current.

Why does the pattern formed by the filings not depend on the direction of the current?

Could iron filings be used to determine the direction that current flows? Why or why not?

Question 8 Functional dependence on current. Select an arbitrary point in space. Vary the current through the wire and record the magnitude of the magnetic field at this point.

I   B

Sketch your data.

B

I

What is the functional dependence of magnetic field on current for a straight, current-carrying wire?

Question 9 Functional dependence on distance. Vary the distance between the point of interest and the wire, and record the magnitude of the magnetic field.

d   B

Sketch your data.

B

d

What is the functional dependence of magnetic field on distance from a straight, current-carrying wire?

Combine your results to express the dependence of magnetic field on both current and distance from the current-carrying wire.

Question 10 Biot–Savart law. What is the distance from a wire carrying +10 A beyond which the magnetic field is less than 15 µT?

Question 11 Biot–Savart puzzler. A 2-cm-long object is placed in the magnetic field of a 15-A wire. One end of the object is exposed to a 35-µT field. What range of magnetic fields may the other end of the object be exposed to?

At what distance is the magnetic field of a 15-A wire equal to 35 µT?

Given your previous answer, at what minimum and maximum distance can the other end of the object be located?

Given this range of positions, compute the range of magnetic fields to which this object may be exposed.

Question 1 Magnetic field lines. What will the magnetic field look like when positive current flows through the loop? (Positive current is defined to flow out of the screen at the top of the loop and into the screen at the bottom of the loop.) Sketch the field.

Question 2 Direction of the magnetic field along the central axis. Why does the magnetic field point to the right along the central axis when positive current flows through the loop? Explain using the right-hand rule.

Question 3 Orientation of the magnetic field. What angle does the magnetic field make relative to the central axis?

Question 4 Magnitude along the central axis. Does the magnitude of the field change along the central axis? If so, describe the change.

Where is the maximum field located?

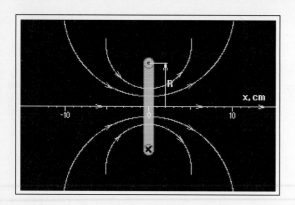

Question 5 Dependence on current. What happens to the magnitude and direction of the magnetic field, at the point in space you are examining, if the current is increased?

Question 6 Flipping the current. What happens to the magnitude and direction of the magnetic field, at the point in space you are examining, if the current is flipped?

Question 7 Iron filings pattern. What happens to the iron filings pattern if the current is flipped between positive and negative values?

Question 8 Functional dependence on current. Select an arbitrary point on the central axis. Vary the current through the loop and record the magnitude of the magnetic field at this point.

I   B

Sketch your data.

B

                                                    I

What is the functional dependence of the magnetic field along the central axis on current?

Question 9 Functional dependence on distance. Vary the distance between the point of interest and the center of the loop, along the central axis, and record the magnitude of the magnetic field.

d   B

Sketch your data.

B

                                                    d

Does there appear to be a simple functional dependence of the magnetic field along the central axis on distance from the loop? If so, what is it?

13.2 Magnetic Field of a Loop continued

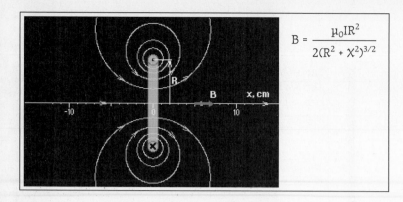

$$B = \frac{\mu_0 I R^2}{2(R^2 + X^2)^{3/2}}$$

Question 10 Finding the maximum field. What is the maximum field along the central axis of a loop carrying −5 A?

Question 11 Using the relationship. What is the distance along the central axis from a loop carrying 10 A beyond which the magnetic field is less that 100 µT?

Question 1 Magnetic field lines. What will the magnetic field look like when positive current flows through the solenoid? (Positive current is defined to flow out of the screen at the top of the loop and into the screen at the bottom of the loop.) Sketch the field.

ⓍⓍⓍⓍⓍⓍⓍⓍⓍⓍⓍⓍⓍⓍⓍⓍⓍⓍⓍⓍⓍ

Question 3 Orientation of the magnetic field. What angle does the magnetic field make relative to the central axis?

Question 2 Direction of the magnetic field inside the solenoid. Why does the magnetic field point to the right inside the solenoid when positive current flows through the loop? Explain using the right-hand rule.

Question 4 Magnitude along the central axis. Does the magnitude of the field change along the central axis?

If so, does it change as rapidly as the field of a single loop?

13.3 Magnetic Field of a Solenoid continued

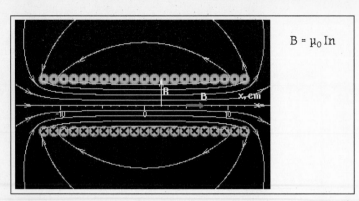

$$B = \mu_0 I n$$

Question 5 Uniformity. What is the magnetic field at the center of the solenoid?

Record the value of the magnetic field at various locations along the central axis.

d	B

Over what portion of the length of the solenoid is the field within 10% of the field at the center? Express this length as a percentage of the entire length of the solenoid.

Question 6 Testing the model. How many loops comprise the solenoid in the simulation?

How long is the solenoid?

What is n, the number of loops per unit length?

Does the theoretical value for the magnetic field of an infinitely long solenoid agree with the simulation?

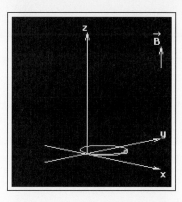

Question 1 Positive or negative? When the particle is initially launched in the +x-direction, what is the direction of the magnetic force on the particle?

Given the directions of the magnetic field, initial magnetic force, and initial velocity, use the right-hand rule to deduce the sign of the electric charge of the particle.

Question 2 Flipping the magnetic field. If the magnetic field is flipped, will the electron traverse its orbit *clockwise* or *counterclockwise*, when viewed from the +z-axis?

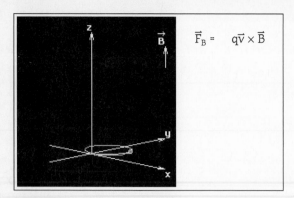

$$\vec{F}_B = \quad q\vec{v} \times \vec{B}$$

Question 3 Increasing the magnetic field: Radius. What happens to the force on the electron if the magnetic field is increased?

Therefore, what will happen to the radius of the electron's path if the magnetic field is increased in magnitude?

Question 4 Increasing the magnetic field: Time. What happens to the electron's velocity if the magnetic field is increased?

From Question 3, what happened to the radius of the electron's path when the magnetic field was increased?

Therefore, what will happen to the time it takes for the electron to complete one orbit if the magnetic field is increased in magnitude?

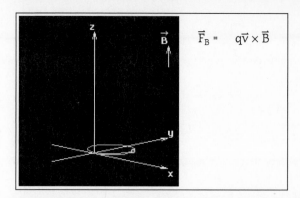

$$\vec{F}_B = q\vec{v} \times \vec{B}$$

Question 5 Increasing the velocity: Radius. What is the relationship between net force and velocity for an object undergoing circular motion?

What happens to the magnetic force on the electron if its velocity is doubled?

Given the changes to both the net force and velocity in the circular motion relationship, what happens to the radius of the electron's path if the initial x-velocity is doubled?

Question 6 Increasing the velocity: Time. From Question 5, what happened to the radius of the electron's path when the velocity was doubled?

Given the change in orbit radius and the change in electron velocity, what will happen to the time it takes for the electron to complete one orbit if the initial x-velocity is doubled?

13.4 Magnetic Force on a Particle continued

$$\vec{F}_B = q\vec{v} \times \vec{B} \qquad\qquad F_{radial} = m\frac{v^2}{r}$$

For Questions 7–9, the initial x-velocity is 4.0×10^7 m/s and the applied magnetic field is 1.0 mT.

Question 7 Varying the field and the velocity I. What happens to the radius and period of the orbit if the x-velocity of the particle is doubled and the magnetic field is doubled?

Explain the change in radius.	Explain the change in period.

Question 8 Varying the field and the velocity II. What happens to the radius and period of the orbit if the x-velocity of the particle is halved and the magnetic field is doubled?

Explain the change in radius.	Explain the change in period.

Question 9 Varying the field and the velocity III. What happens to the radius and period of the orbit if the x-velocity of the particle is doubled and the magnetic field is halved?

Explain the change in radius.	Explain the change in period.

Question 10 z-Velocity. Does the path of the particle change if it is given an additional velocity component in the z-direction?

Sketch your prediction for the electron's path.

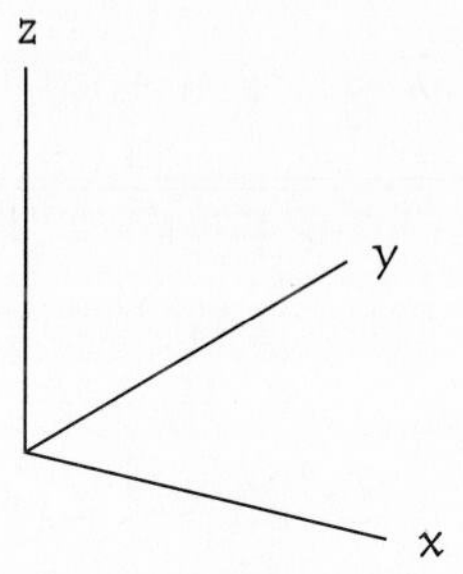

Prediction

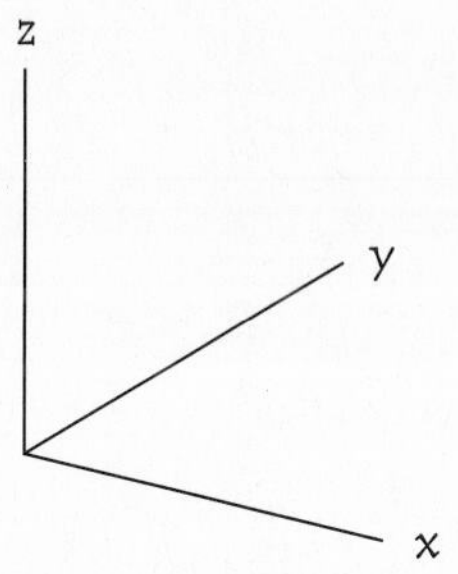

Actual

Question 11 Exclusively z-velocity. Predict the resulting motion when the particle is launched along the z-axis.

Question 1 Field produced by wire 1. If a positive (directed out of the screen) current flows through the left wire, is the direction of the magnetic field produced at the location of the right wire *up, down, left,* or *right?* Sketch the field due to wire 1.

⊙   ⊙
1   2

Question 3 Field produced by wire 2. Since a positive (directed out of the screen) current flows through the right wire, is the direction of the magnetic field produced at the location of the left wire *up, down, left,* or *right?* Sketch the field due to wire 2.

⊙   ⊙
1   2

Question 2 Force on wire 2. If a positive current flows through the right wire, does the magnetic field produced by the left wire exert a force directed *up, down, left,* or *right* on the right wire? Sketch the force on wire 2.

⊙   ⊙
1   2

Question 4. Force on wire 1. Since a positive current flows through the left wire, does the magnetic field produced by the right wire exert a force directed *up, down, left,* or *right* on the left wire? Sketch the force on wire 1.

⊙   ⊙
1   2

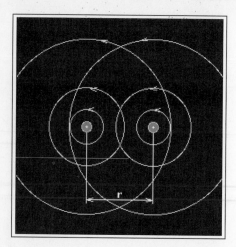

Question 5 Separation dependence.
What happens to the magnitude of the magnetic field at the location of each wire when the wires are brought closer together?

What happens to the magnitudes of the magnetic forces acting on the wires when the wires are brought closer together?

Question 6 Reversing current.
What happens to the directions of the magnetic forces acting on the wires when the current through wire 1 is reversed?

Does it matter whether current 1 or current 2 is reversed?

Question 7 Unequal forces. Is it possible, by varying the currents and/or the separation, for the forces of interaction between the two parallel wires to be unequal?

If so, what combination of currents and/or separations accomplishes this task?

Question 8 Remembering mechanics. What law of mechanics is illustrated by the preceding question?

©1999 Addison Wesley Longman, Alan Van Heuvelen and Paul D'Alessandris

$$\vec{F}_B = I\vec{L} \times \vec{B}$$

$$B = \frac{\mu_0 I}{2\pi r}$$

Question 9 Calculating the field. What is the magnetic field at the location of each wire in a pair of wires, one carrying 2.0 A and the other −1.2 A, 0.5 m apart?

Magnetic field at 2.0-A wire, due to −1.2-A wire:	Magnetic field at −1.2-A wire, due to 2.0-A wire:

Question 10 Calculating the force per unit length. What is the force per unit length acting on each wire in a pair of wires, one carrying 2.0 A and the other −1.2 A, 0.5 m apart?

What is the force per unit length acting on the 2.0-A wire, due to the −1.2-A wire?	What is the force per unit length acting on the −1.2-A wire, due to the 2.0-A wire?

Question 11 Varying the current. What happens to the force between the wires if both the currents are halved?

What happens to the magnetic fields produced?	What happens to the force between the wires?

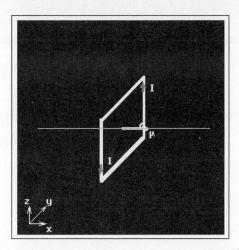

Question 1 Magnetic force on the top of the loop. What is the direction of the current flowing through the wire comprising the top of the loop?

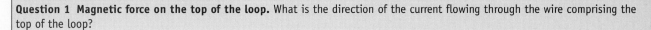

If a magnetic field is applied in the positive x-direction, what is the direction of the magnetic force on the wire comprising the top of the loop?

Question 2 Reversing the field. If the direction of the magnetic field is reversed, what happens to the direction of the magnetic force on each of the four sides of the loop?

13.6 Magnetic Torque on a Loop continued

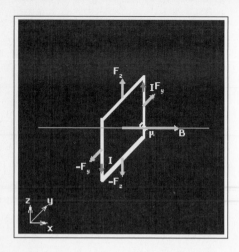

Question 3 Net force on the loop. What is the net force acting on the loop?

Question 4 Rotating the loop: Forces. What happens to the magnitude and direction of the magnetic forces acting on the four wire segments if the loop is rotated to a small positive angle?

Top wire:

Bottom wire:

Left wire:

Right wire:

©1999 Addison Wesley Longman, Alan Van Heuvelen and Paul D'Alessandris

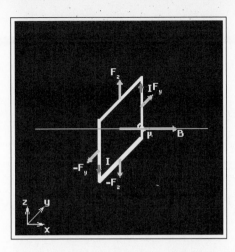

Question 5 Rotating the loop: 90°.

What is the direction of current flow through the top wire when the loop is rotated to +90°?	What happens to the magnitude of the force on the top wire when the loop is rotated to +90°?

Question 6 Rotating the loop: Beyond 90°. What happens to the magnitude and direction of the force on the top wire when the loop is rotated beyond +90°?

Question 7 Rotating the loop: Net force. Can the net force on the loop ever be non-zero? Explain.

13.6 Magnetic Torque on a Loop continued

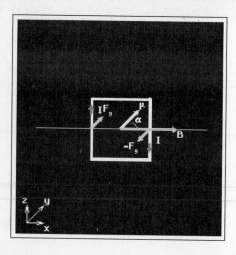

Question 8 Rotating the loop: Torque. What is the direction of the net torque on the loop if it is rotated to a small positive angle?

Question 9 Rotating the loop: Negative angles. What is the direction of the net torque on the loop if it is rotated to a small negative angle?

Question 10 Maximum positive torque. With the magnetic field pointing in the +x-direction, for what angle of orientation does the loop experience the maximum positive torque?

Question 11 Dependence on area. Does the size of the torque depend on the area of the loop? If so, how?

Question 12 General rule. For various orientations, closely examine the relationship between the magnetic field, magnetic dipole moment, and the torque vectors. Measure the torque for various angles between the magnetic field and the dipole moment, and sketch the results.

θ	τ
−180°	
−135°	
−90°	
−45°	
0°	
45°	
90°	
135°	
180°	

τ

θ

Can you think of a simple vector relationship that summarizes the dependence of the torque on the magnetic dipole moment and the magnetic field?

Question 1 Trajectory. Will the trajectories of the two isotopes be the same or different? Explain.

Sketch the trajectories.

Question 2 Doubling the magnetic field. What happens to the radius of curvature of the trajectories if the magnitude of the magnetic field is doubled? Explain.

Sketch the trajectories.

Question 3 Doubling the velocity. What happens to the radius of curvature of the trajectories if the magnitude of the velocity is doubled? How does this radius compare to the original radius before the magnetic field and the velocity were doubled? Explain.

Sketch the trajectories.

Question 4 Relating curvature to velocity and magnetic field. Using your knowledge of the magnetic force and the mechanics of a particle exhibiting circular motion, derive a relationship between the radius of the particle's trajectory, the particle's velocity, and the magnetic field.

(a) Relation for the acceleration of an object undergoing circular motion:

(b) Relation for the magnetic force on a charged particle:

Combine the two relations using Newton's second law and solve for the radius of the circular orbit.

Question 5 Predicting the curvature. Using the result derived in Question 4, calculate the radius of curvature for both isotopes of carbon. (Remember that the isotopes are singly ionized and that one atomic mass unit is 1.66×10^{-27} kg.)

Question 6 Adjusting the velocity. With what velocity would the neon beam need to be injected into the mass spectrometer for ^{20}Ne to have a radius of curvature equal to that of the previous ^{12}C beam?

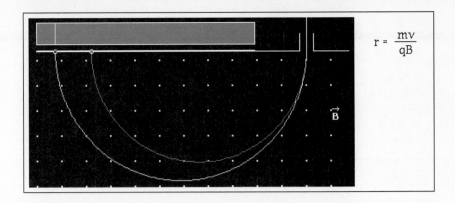

$$r = \frac{mv}{qB}$$

Question 7 Maximum separation for carbon. What is the maximum separation between the two isotopes of carbon?

Parameter settings:

B =

v =

Data:

r_1 =

r_2 =

$r_2 - r_1$ =

Question 8 Maximum separation for neon. Do you think that the maximum separation for neon will be larger or smaller than that for carbon? Is neon harder or easier to separate into its constituent isotopes than carbon? Explain.

Parameter settings:

B =

v =

Data:

r_1 =

r_2 =

$r_2 - r_1$ =

Question 9 Maximum separation for uranium. Is uranium harder or easier to separate into its constituent isotopes than neon? Explain.

Parameter settings:

B =

v =

Data:

r_1 =

r_2 =

$r_2 - r_1$ =

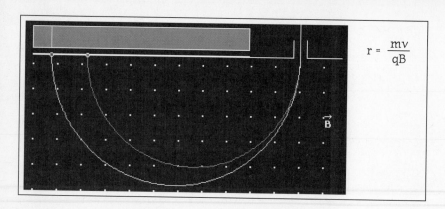

$$r = \frac{mv}{qB}$$

Question 10 Unknown element. Determine the masses of the two isotopes of the unknown element. From your knowledge of chemistry, determine the identity of the unknown element.

Smaller radius:

Larger radius:

Mass of smaller-mass isotope in kg:

Mass of larger-mass isotope in kg:

Mass of smaller-mass isotope in amu:

Mass of larger-mass isotope in amu:

Unknown element:

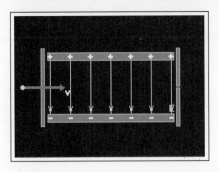

Question 1 Electric force on an electron. If an electron is fired into the region between the charged parallel plates, will the electron be deflected upward or downward by the electric force?

Question 3 Magnetic force on an electron. If an electron is fired into the region between the parallel plates, should the magnetic field be oriented into or out of the screen to create a magnetic force downward?

Question 2 Increasing the electric field. If the magnitude of the electric field is increased, what happens to the curvature of the electron's path between the parallel plates?

Question 4 Increasing the magnetic field. If the magnitude of the magnetic field is increased, what happens to the curvature of the electron's path between the parallel plates?

Question 5 Adjusting the magnetic field. Does the electron deflect up or down?

Therefore, which force is larger?

How should the magnetic field be adjusted for the electron to pass through the device undeflected?

Question 6 Adjusting the electric field. Does the electron deflect up or down?

Therefore, which force is larger?

How should the electric field be adjusted for the electron to pass through the device undeflected?

Question 7 Adjusting the velocity. Does the electron deflect up or down?

Therefore, which force is larger?

How should the velocity of the electron be adjusted for it to pass through the device undeflected?

Question 8 Flipping the electric field. What happens to the electron's path if the direction of the electric field is flipped? Will it still pass through the device undeflected?

Question 9 Flipping the magnetic field. What happens to the electron's path if the direction of the magnetic field is now flipped? Will the electron pass through the device undeflected?

Question 10 Opposite charge. If the sign of the charge of the incident particle is flipped, what happens to the electric force?

Magnitude: Direction:

What happens to the magnetic force?

Magnitude: Direction:

Given that an electron, with negative charge, passes through the device undeflected, what must be changed so that a positron, with positive charge, will pass through undeflected?

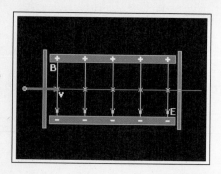

Question 11 Larger mass. If the mass of the incident particle is increased, what happens to the electric force?

Magnitude: Direction:

What happens to the magnetic force?

Magnitude: Direction:

Given that a positron passes through the device undeflected, what must be changed so that a proton, with larger mass, will pass through undeflected?

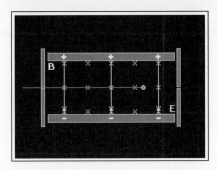

Question 12 Larger charge. If the charge of the incident particle is increased, what happens to the electric force?

Magnitude: Direction:

What happens to the magnetic force?

Magnitude: Direction:

Given that a proton passes through the device undeflected, what must be changed so that a helium nucleus, with twice the electric charge, will pass through undeflected?

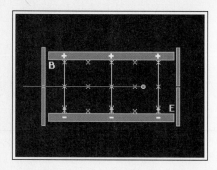

Question 13 Determining the velocity. If a particle passes through a selector undeflected, the magnitudes of the electric and magnetic forces must be equal. Using this observation, determine a relationship between the velocity of undeflected particles and the electric and magnetic fields present in the device.

Relation for magnetic force on a charge q:

Relation for electric force on a charge q:

Set the electric and magnetic forces equal and solve for the velocity.

Question 14 What about neutrons? At what value(s) of velocity, magnetic field, and electric field will a neutron pass through the velocity selector undeflected?

Question 15 Atomic velocities. Is a velocity selector an effective apparatus for determining the velocity of a beam of atoms?

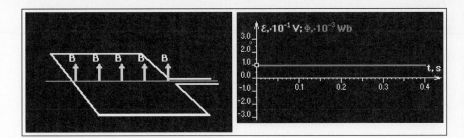

Question 1 Magnetic flux: Magnetic field magnitude.
What happens to the magnetic flux through the loop when the magnetic field is increased to positive values?

Question 3 Magnetic flux: Loop area. What happens to the magnetic flux through the loop when the area of the loop is decreased?

Question 2 Magnetic flux: Magnetic field orientation.
What happens to the magnetic flux through the loop when the magnetic field is flipped to negative values?

Question 4 Magnetic flux: Loop orientation. What happens to the magnetic flux through the loop as the loop rotates relative to the magnetic field? What is the value of the flux after the loop has rotated through 90°? Through 180°?

Question 5 Magnetic flux: Summary. Summarize the dependence of magnetic flux on the magnetic field, the area of the loop, and the relative orientation between them.

13.9 Electromagnetic Induction continued

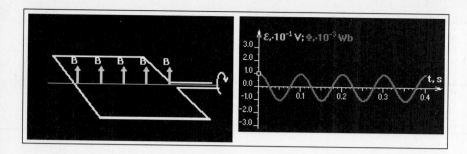

Question 6 Constant flux. Will an EMF be induced when the rotational frequency of the loop is zero?

Question 8 Maximum induced EMF. For what loop orientation(s) is the induced EMF a maximum in magnitude?

Question 7 Rotating the loop. Will an EMF be induced when the rotational frequency of the loop is non-zero?

Question 9 Maximum rate of change of flux. For what loop orientation(s) is the magnetic flux changing most rapidly?

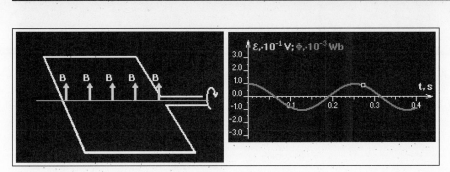

Question 10 No induced EMF. For what loop orientation(s) is the induced EMF instantaneously zero?

Question 11 Zero rate of change of flux. For what loop orientation(s) is the magnetic flux instantaneously not changing?

Question 12 Relating EMF to magnetic flux. Based on your observations, postulate a possible relationship between the induced EMF and the rate of change of the magnetic flux.

Question 13 Dependence of induced EMF on frequency. As the frequency of rotation increases, what happens to the maximum induced EMF?

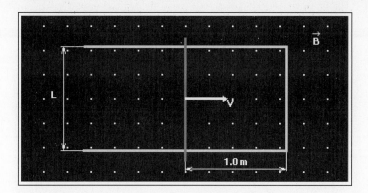

Question 1 Magnetic flux: Stationary bar.

What is the area of the loop?

What is the magnetic flux through the loop?

Question 2 Magnetic flux: Bar moving to the right. What happens to the magnetic flux through the loop as the conducting bar moves to the right?

Question 3 Magnetic flux: Bar moving to the left. What happens to the magnetic flux through the loop as the conducting bar moves to the left?

Question 4 Change in magnetic flux. Complete the following table.

V	$\Phi_{initial}$	Φ_{final}	$\Delta\Phi$
+10			
+7			
+3			
−3			
−7			
−10			

Does the change in magnetic flux depend on the velocity of the bar? Explain.

Question 5 Rate of change of magnetic flux. Complete the following table.

V	$\Delta\Phi$	t	$\Delta\Phi/t$
+10			
+7			
+3			
−3			
−7			
−10			

Does the rate of change of magnetic flux depend on the velocity of the bar? Explain.

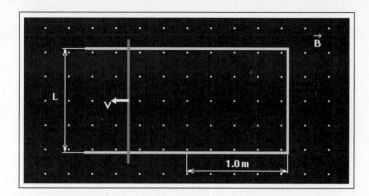

Question 6 Rate of change of area. Construct a relationship between the rate of change of area of the loop and the bar's length and velocity. What is the distance the bar travels in a time Δt?

What is the change in the area of the loop in a time Δt? (Be careful with signs.)

What is the rate of change of the area of the loop?

Question 7 Rate of change of flux. Given the previous results, construct a relationship between the rate of change of magnetic flux through the loop and other relevant variables.

Question 8 Induced EMF. Given the previous result, construct a relationship between the induced EMF in the moving bar and relevant variables.

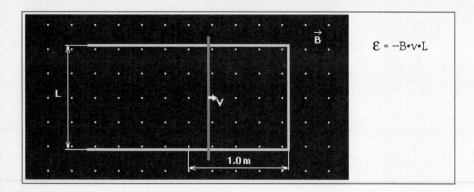

$$\mathcal{E} = -B \cdot v \cdot L$$

Question 9 Testing the relation. Complete the following table.

B	V	L	EMF
0.1	+10	1.0	
0.1	−5	0.5	
0.1	0	1.0	
0	+10	0.5	
−0.05	−10	0.2	
−0.05	+5	0.5	
−0.05	+10	0.2	

Is the EMF induced in the moving bar always in agreement with the derived relationship?

Question 10 Direction of the induced current. What is the sign of the magnetic flux?

How is the magnetic flux changing as the bar moves?

Will the induced EMF drive current to produce positive or negative magnetic flux?

Which way will the induced current flow in the moving bar?

Question 11 Direction of the induced current, again. What is the sign of the magnetic flux?

How is the magnetic flux changing as the bar moves?

Will the induced EMF drive current to produce positive or negative magnetic flux?

Which way will the induced current flow in the moving bar?

Question 12 Direction of the induced current, one last time. What is the sign of the magnetic flux?

How is the magnetic flux changing as the bar moves?

Will the induced EMF drive current to produce positive or negative magnetic flux?

Which way will the induced current flow in the moving bar?

Question 13 Maximum current. What combination of parameters will produce the maximum magnitude current? What should B equal, and why?

What should v equal, and why?

What should L equal, and why?

What should R equal, and why?

ac CIRCUITS

14

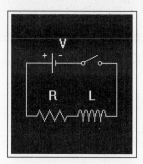

Question 1 A simple circuit. Imagine a circuit consisting of a 2-Ω resistor, a 4-V battery, and a switch. (**Note:** There is no inductor [L = 0] in this hypothetical circuit.) What current will flow through the circuit when the switch is closed?

Question 2 Faraday's law of induction. State as clearly as you can, without using equations, Faraday's law of induction.

Question 3 Closing the switch. When the switch is closed, will the EMF induced in the inductor act in the same direction or the opposite direction of the EMF of the battery? Why?

Question 4 Much later. What current flows through the circuit long after the switch is closed? Why?

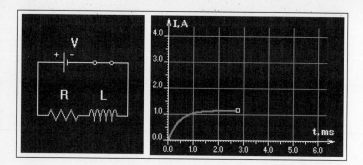

Question 5 Doubling the inductance. When the inductance is doubled, does the time for the current to reach its equilibrium value *increase, decrease,* or *stay the same?* Why?

Question 6 Doubling the resistance. When the resistance is doubled, does the time for the current to reach its equilibrium value *increase, decrease,* or *stay the same?* Why?

Question 7 Inductive time constant—I. With the resistance set to 1.5 Ω, the inductance to 2.5 mH, and the potential difference to 3.0 V, predict the equilibrium current and the inductive time constant.

$I_{equilibrium}$:

Time constant, τ:

Question 8 Inductive time constant—II. Predict the value of the current at $t = \tau_L$ and $t = 2\tau_L$.

$I(\tau_L) =$

$I(2\tau_L) =$

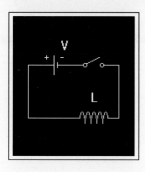

Question 1 Current vs. time. What behavior does the current through the inductor exhibit?

Sketch a graph of I vs. t.

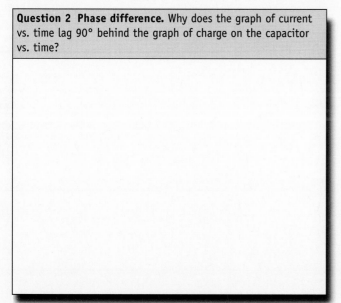

Question 2 Phase difference. Why does the graph of current vs. time lag 90° behind the graph of charge on the capacitor vs. time?

Question 3 Initial energy. At t = 0 s, is the energy in the circuit in the inductor or the capacitor? Why?

Question 4 Energy flow. After the switch is closed, where does the energy initially stored in the capacitor go?

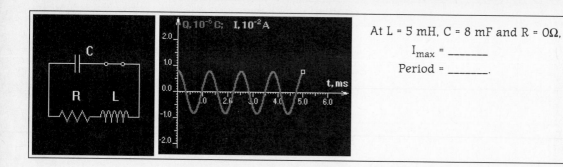

At L = 5 mH, C = 8 mF and R = 0Ω,

$$I_{max} = \underline{\hspace{2cm}}$$

$$Period = \underline{\hspace{2cm}}.$$

Question 5 Double inductance. What happens to the maximum current through the circuit when the inductance is doubled? Why?

What happens to the period of oscillation when the inductance is doubled? Why?

Question 6 Halve capacitance. What happens to the maximum current through the circuit when the capacitance is halved? Why?

What happens to the period of oscillation when the capacitance is halved? Why?

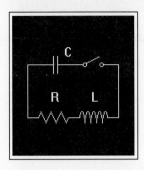

Question 7 Adding resistance. What happens to the behavior of the circuit if resistance is added?

Sketch a graph of I vs. t.

I

t

Question 8 Energy loss? Where does the energy go in a circuit that includes resistance?

Question 9 Mechanical analogy—I. A large inductance makes the circuit respond less rapidly to outside "forces." What aspect of a spring–mass system acts in an analogous way? Why?

Question 10 Mechanical analogy—II. A large capacitance allows the storage of a *large* amount of energy through the application of a relatively *small* potential. Does some aspect of a spring–mass system act in an analogous way? Why?

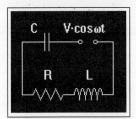

Question 1 Potential difference across the resistor. When attached to a source of alternating electric potential difference, does the current through a purely resistive element lead the potential difference across it by 90°, lag behind the potential difference by 90°, or is it in phase with the potential difference?

Sketch the current in correct phase relation to the given potential difference.

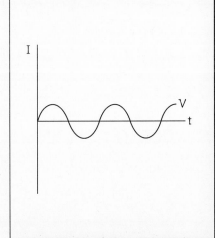

Question 2 Potential difference across the inductor. When attached to a source of alternating electric potential difference, does the current through an inductor lead the potential difference across it by 90°, lag behind the potential difference by 90°, or is it in phase with the potential?

Sketch the current in correct phase relation to the given potential difference.

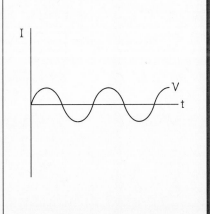

Question 3 Potential difference across the capacitor. When attached to a source of alternating electric potential difference, does the current "through" a capacitor lead the potential difference across it by 90°, lag behind the potential difference by 90°, or is it in phase with the potential difference?

Sketch the current in correct phase relation to the given potential difference.

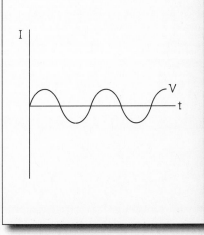

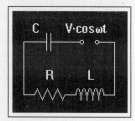

Question 4 X_L. Suppose the driving frequency, ω, is increased. What happens to the rate of change of current through the inductor?

What happens to the potential difference across the inductor?

What happens to the reactance of the inductor?

Question 5 X_C. Now the driving frequency, ω, is increased. What effect does this have on the amount of charge that has time to build up on the capacitor?

What happens to the potential difference across the capacitor?

What happens to the reactance of the capacitor?

Question 6 Phasor diagram. With the driving frequency set to 1400 rad/s, sketch a phasor diagram for the circuit. Set the phasor representing the resistor on the horizontal axis.

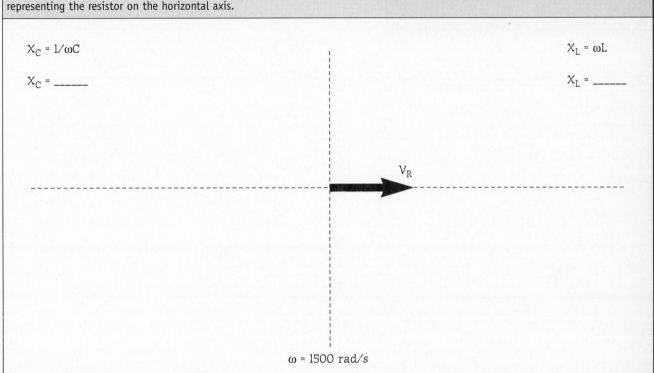

$X_C = 1/\omega C$

$X_C =$ _____

$X_L = \omega L$

$X_L =$ _____

V_R

$\omega = 1500$ rad/s

Question 7 Increasing ω. What happens to the phasor diagram as the angular frequency is increased from 1500 rad/s to 2500 rad/s?

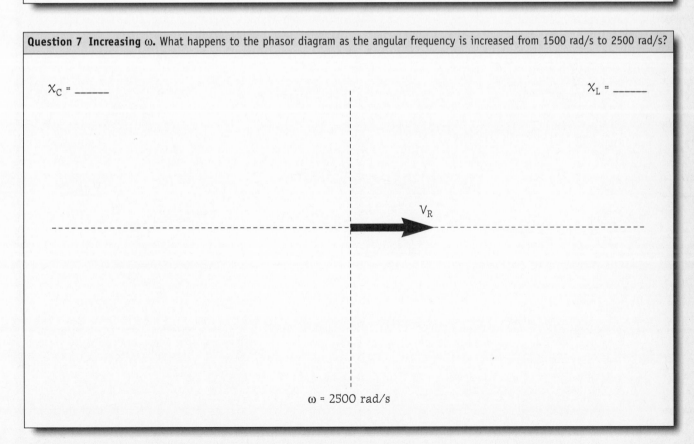

$X_C =$ _____

$X_L =$ _____

V_R

$\omega = 2500$ rad/s

14.3 The Driven Oscillator continued

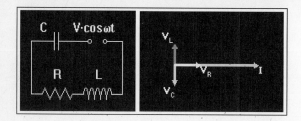

Question 8 Resonance. At what driving frequency will the circuit be at resonance?

Question 9 Resonance curve. How does the shape of the resonance curve change if the resistance of the circuit is reduced? Sketch the resonance curve for R < 3 Ω.

$R = 3.0\ \Omega$

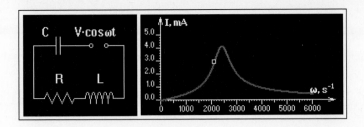

Question 10 Phasor practice. Sketch the phasor diagram for a circuit with a 3.0-mH inductor, a 100-uF capacitor, and driven at 4000rad/sec.

X_C = _____

X_L = _____

V_R

Question 11 Resonance practice. Calculate the resonance frequency for the circuit described in Question 10.

GEOMETRIC OBJECTS

15

Question 1 Speed of light. The index of refraction of water is about 1.35. Determine the speed of light in water.

Question 2 Reflected angle. Set the refractive index of the lower medium to 1.35. Try different incident angles and observe the reflected angles. State a rule that relates the incident and reflected angles. This rule applies to smooth surfaces.

Question 3 Snell's law and the refracted angle. Leave the refractive index of the upper medium at 1.00 and of the lower medium at 1.50. Try different incident angles and observe the refracted angles. Try different incident angles and observe the refracted angles. Show that the incident angle θ_1 and the refracted angle θ_2 are related by Snell's law, $n_1 \sin \theta_1 = n_2 \sin \theta_2$, for two examples where the incident ray moves from the smaller to the larger index of refraction material. Draw each situation before trying the calculations.

Question 4 Snell's law and the refracted angle. Leave the index of refraction of the upper medium at 1.00 and change the index of refraction of the lower medium to 1.60. Choose the incident angle for one example showing that Snell's law also applies when the incident ray moves from the larger to smaller index of refraction material. Draw each situation before trying the calculations.

Question 5 Determine the direction of the refracted ray for light in air incident at 45° on an interface with 1.55 refractive index glass. Draw each situation before trying the calculations.

Question 6 Determine the direction of the refracted ray for light in water (n = 1.35) incident at 30° on a surface with air. Draw the situation before trying the calculations.

Question 1 Total internal reflection. Draw several rays for different incident angles showing how the direction the wave is traveling changes as it moves from the larger refractive index medium at the bottom into the smaller refractive index medium at the top. What is the direction of the refracted wave for the incident critical angle θ_{crit}?

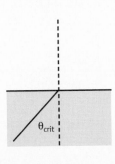

Question 2 Moving from a smaller to larger refractive index medium. Draw several rays for different incident angles showing how the direction the wave is traveling changes as it moves from the small refractive index medium at the top into the large refractive index medium at the bottom. What condition seems to be necessary for total internal refraction to occur?

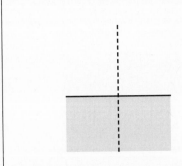

Question 3 Use Snell's law, $n_1 \sin \theta_1 = n_2 \sin \theta_2$, to determine the critical angle for light passing from the lower medium with refractive index 1.60 into the upper medium with refractive index 1.35. Indicate the indices of refraction of the two media and draw the ray representing the direction the wave is traveling at this critical angle before and after reaching the interface between the two media.

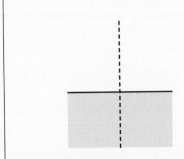

Question 4 Total internal reflection. Use Snell's law to develop a general rule for determining the critical angle θ_{crit} for light passing from the large refractive index medium n_1 (bottom) into the upper small refractive index medium n_2 (top).

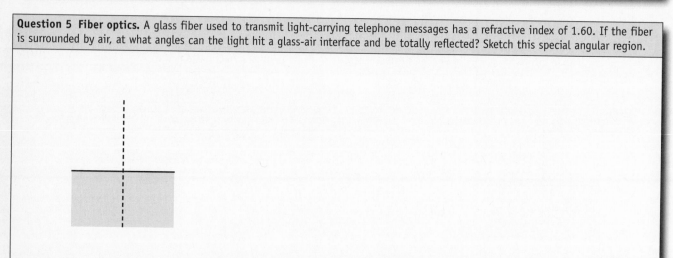

Question 5 Fiber optics. A glass fiber used to transmit light-carrying telephone messages has a refractive index of 1.60. If the fiber is surrounded by air, at what angles can the light hit a glass-air interface and be totally reflected? Sketch this special angular region.

Question 6 A coated-fiber optics problem. A glass fiber used to transmit light-carrying telephone messages has a refractive index of 1.60. This fiber is surrounded by a coating with refractive index 1.35. The coating is surrounded by air of refractive index 1.00. At what angles can the light hit a glass-coating interface and not escape the coated fiber into the air? Sketch this special angular region.

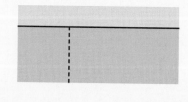

Question 1 Suppose light travels from air with refractive index 1.00 into glass with refractive index 1.55. If the incident angle is 30°, determine the angle of the refracted light as it passes through the glass.

Question 2 Suppose light travels from glasslike material with refractive index 1.55 into water with refractive index 1.35. If the refracted angle as the light passes through the water is 37°, determine the angle of the incident light coming from the glass.

Question 3 Suppose you are swimming underwater in a lake with a smooth surface and a flat shoreline (see the figure). While underwater, you see a friend sunbathing on the shore. At what angle are you looking? The index of refraction is about 1.35 for water and 1.00 for air.

Question 4 Suppose that while swimming in water of refractive index 1.35, your keys fall out of your bathing suit. After drying off at the poolside, you see them in the water and decide to retrieve them with a bow and arrow with a magnetic tip. Your eyes are 1.5 m above the water's surface and the pool is 1.5 m deep (see figure). When you look at an angle of 40° below the horizontal, you see the keys. At what angle should you aim the bow, assuming that the arrow does not change direction when entering and traveling through the water?

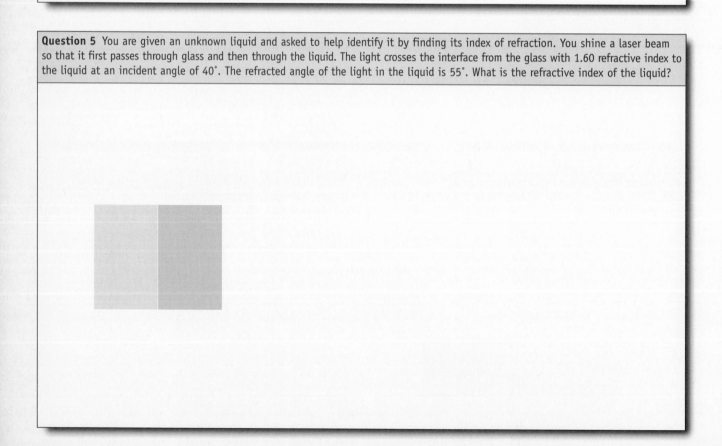

Question 5 You are given an unknown liquid and asked to help identify it by finding its index of refraction. You shine a laser beam so that it first passes through glass and then through the liquid. The light crosses the interface from the glass with 1.60 refractive index to the liquid at an incident angle of 40°. The refracted angle of the light in the liquid is 55°. What is the refractive index of the liquid?

Question 1 Suppose an arrowhead is located 1.00 m (y = 1.00 m) from a plane mirror above the left edge of the mirror (x = 0.00 m). Your eye is about 1.0 m from the mirror and above the middle of the mirror. Draw three rays representing light that travels from the arrowhead toward the mirror and then after reflection toward the region where your eye is located—the rays point in slightly different directions. Determine the point from which the light seems to originate. This is the point where the image of the arrowhead in the mirror is located.

Question 2 Other object locations. Grab the arrowhead with the pointer and move it to different locations. Observe the image location for different object locations. Then, construct a general rule for the location of the image of an object as seen in a plane mirror.

Question 3 Taking your picture in a mirror. Suppose you are standing 2.0 m in front of a mirror and wish to take your picture by aiming a camera at your image in the mirror. What distance setting should you have on your camera lens? Explain.

Question 1 Focal length and radius of curvature. Move the focal length f of the mirror to different values and observe the radius of curvature slider R of the mirror. Devise a rule showing how the mirror's focal length f is related to its radius of curvature R. Note the location of the focal point, the blue cross on the axis of the lens. Try both concave and convex mirrors. The focal point is off the screen for some mirrors.

Question 2 Path of the horizontal ray. Consider light that moves horizontally from the tip of the object arrow toward the mirror. Try different focal-length mirrors and devise a rule for the path followed by that ray after reflection from the mirror. Be sure to consider both concave (positive f) and convex (negative f) mirrors.

Question 3 Ray reflected from center of mirror. Consider light that moves from the tip of the object arrow toward the center of the mirror. Try different focal-length mirrors and devise a rule for the path followed by that ray after reflection from the mirror. Be sure to consider both concave and convex mirrors.

Question 4 Virtual-image ray diagram. As you draw the following diagrams, make them cover most of the space given and use the scale of the drawing to estimate the image distance. Draw a ray diagram using our two special rays and estimate the image location for the given object distances.

50-mm object distance from a +100-mm focal-length mirror:

axis

mirror

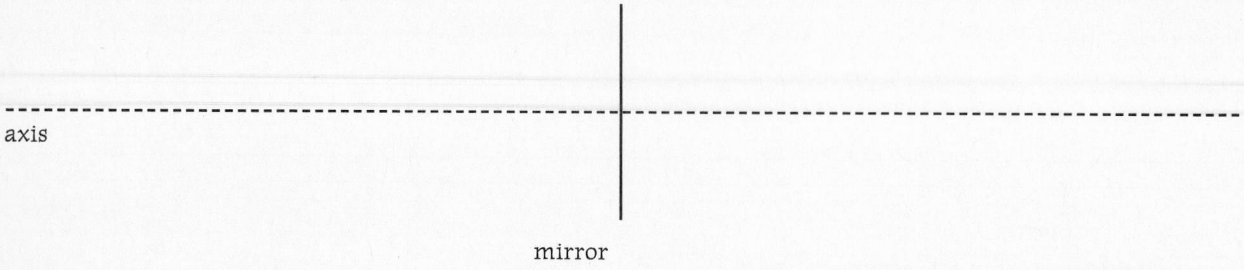

81-mm object distance from a +250-mm focal-length mirror:

axis

mirror

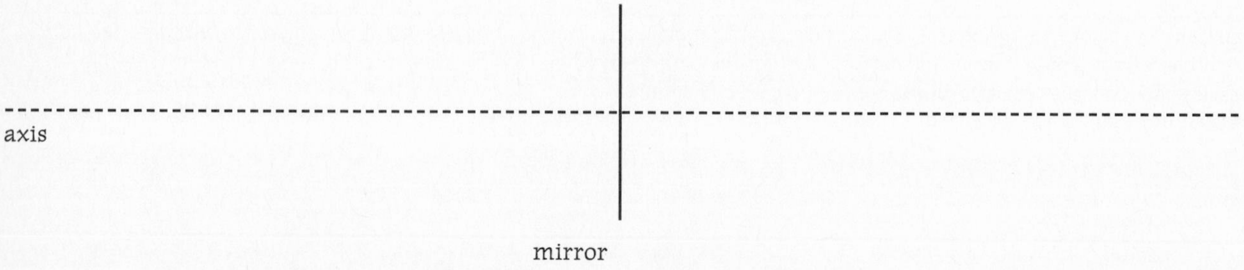

200-mm object distance from a –100-mm focal-length mirror:

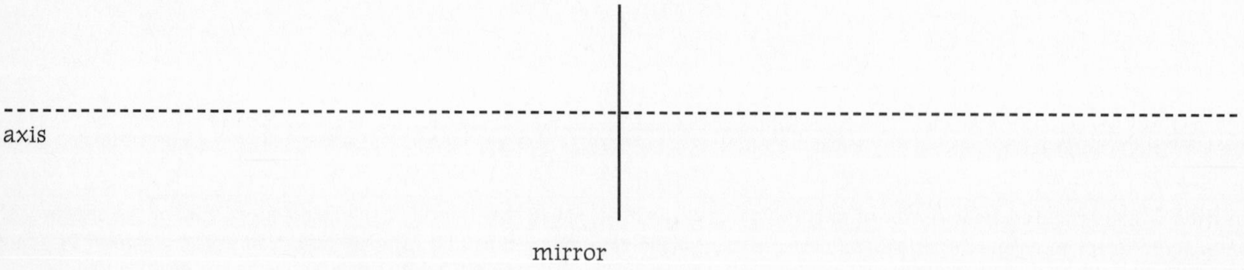

axis

mirror

200-mm object distance from a –250-mm focal-length mirror:

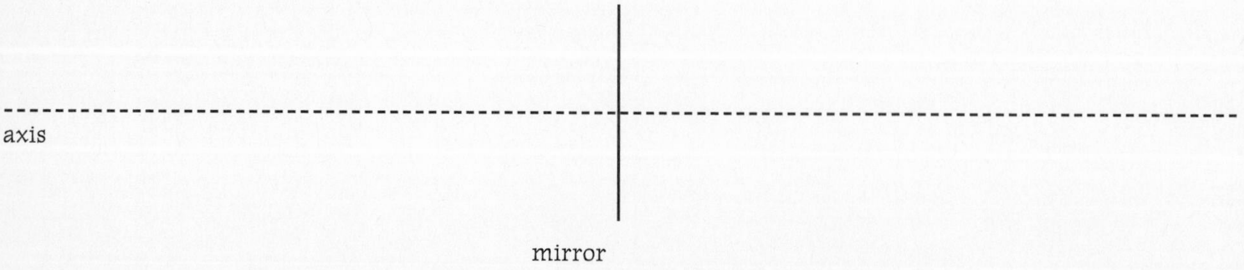

axis

mirror

Question 5 Real-image ray diagram. As you draw the following diagrams, make them cover most of the space given and use the scale of the drawing to estimate the image distance. Draw a ray diagram using our two special rays and estimate the image locations for the given object distances.

250-mm object distance from a +100-mm focal-length mirror:

axis

mirror

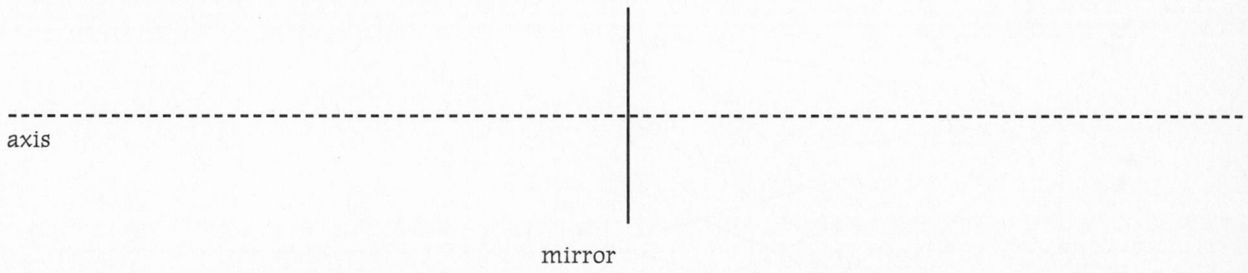

200-mm object distance from a +100-mm focal-length mirror:

axis

mirror

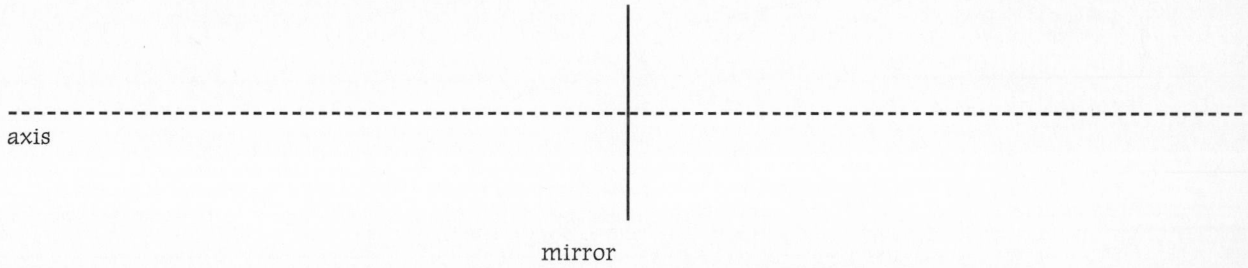

170-mm object distance from a +100-mm focal-length mirror:

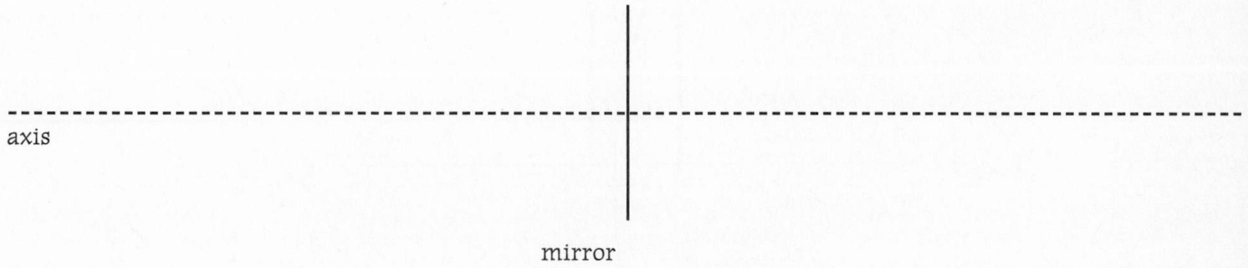

axis

mirror

Question 6 Other rays. The two rays shown in the simulation are convenient for locating mirror images because we can predict their directions if we know the focal point of the mirror. However, many other rays can be drawn to represent the light's direction of travel along other paths (see the figure).

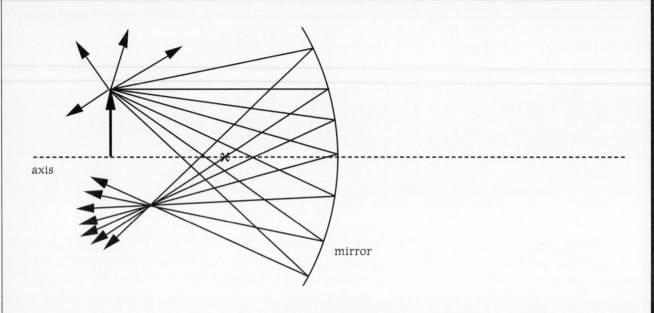

Suppose a cover is placed over the top half of the mirror in the figure as shown here.

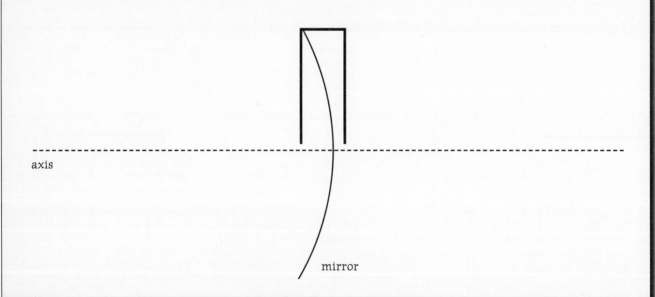

What effect does this have on the image? Explain your reasoning.

Question 1 Use the mirror equation to find the unknown quantity in the following table. You can adjust the simulation to check your answers.

	s (mm)	s' (mm)	f (mm)
(a)	250	_____	100
(b)	50	_____	100
(c)	200	−111	_____
(d)	_____	240	120
(e)	180	_____	100
(f)	_____	−94	−150

Question 2 Surveillance mirrors. Which situation(s) in Question 1 resemble surveillance mirrors used in stores to prevent shoplifting and at the corners of hallways to prevent collisions?

Question 1 Observing image height. Set the focal length of the mirror to f = +100 mm. Move the object so it is the farthest distance from the mirror but can still be seen. Note the relative size of the image and object. Now, with your pointer, move the object toward the mirror. Note carefully the relative sizes of the image and object. Write a qualitative description of your observations concerning these heights. Restrict your experiment to positions that produce observable real images.

Question 2 Linear magnification m. (a) Set the focal length of the mirror to f = +100 mm and the object distance to s = 250 mm. Estimate the ratio of the image height and the object height (h'/h). This is called the linear magnification m. How does your estimate compare to the simulation value given for the linear magnification m? Take the ratio of the image and object positions (s'/s)

(b) Move the object closer to the mirror so that the real image is as far from the mirror as possible but can still be seen. Estimate the ratio of the image and object heights (h'/h) and compare this estimate to the simulation's reported linear magnification m and to the ratio s'/s.

(c) Set the focal length to f = –500 mm and the object position to s = 120 mm. Estimate the ratio of the image height and object height (h'/h). Compare your estimate to the simulation's reported linear magnification m and to the ratio s'/s.

(d) Try focal lengths f = –250, –180, –120, and –100 mm. Change the object distance for each focal length so that the image appears to be half the height of the object and note the value of the linear magnification m and the ratio s'/s for each case.

(e) Finally, state in the form of an equation a rule that relates the ratio of the image and object heights (h'/h), the linear magnification m, and the ratio of the image and object distances s'/s.

Question 3 Why does h'/h = −s'/s? Set f = +100 mm and s = 200 mm. Note that m = h'/h = −s'/s = −1.0. Use the two triangles caused by the ray that reflects from the center of the lens to justify why h'/h = -s'/s (see the figure).

Question 4 Check h'/h = −s'/s for another triangle. Set f = −170 mm and s = 170 mm. Note that m = h'/h = −s'/s = 0.50. Use the two triangles shown below to justify this equality, including the negative sign in front of s'/s.

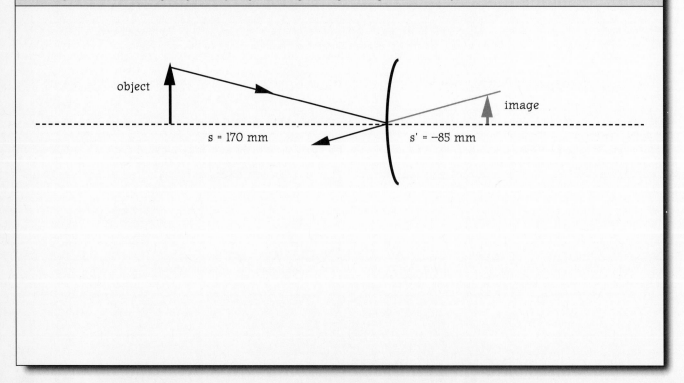

Question 1 Makeup and shaving mirror. A business person asks you to design a mirror that can be used for makeup or shaving. The image formed by the mirror is to be upright and 1.62 times bigger than the object, a user's face. The image is to be 20.9 cm from the object. Determine the focal length needed for the mirror.

axis --

mirror

Question 2 Fortune teller's crystal ball. Your fortune teller looks into his crystal ball whose radius is 200 mm (be careful in choosing the focal-length sign—remember the shape of a crystal ball). The teller's eye is 250 mm from the surface of the ball. Determine the location of the image of his eye and the width of the image of the 3.0-cm-wide eye.

axis --

mirror

Question 3 Surveillance mirror for ants. To help ants avoid collisions when entering their ant hill, you decide to build a surveillance mirror at the entrance. The mirror is convex with a focal length of –100 mm. Where is the image and what is the image height and orientation of a 2.0-mm-tall ant that is 15 cm from the mirror? Repeat your calculations for a 2.0-mm-tall ant that is 5.0 cm from the mirror.

axis --

mirror

15.8 Spherical Mirrors: Problems continued

Question 4 Inverted light hanging in space. You are asked to design a display for a museum that appears to have an inverted glowing light bulb floating in space. Play around with the spherical mirror simulation to see how you might do this. First produce such a display with the simulation (in the form of object and image arrows). Then provide drawings of the situation and an analysis using the mirror equation and the magnification information so that the museum curator is convinced that you know what you are talking about. **Note:** There are many possible answers for this design.

axis

mirror

Question 5 Light beam. You are asked to use a bright light bulb and a large mirror to produce a parallel beam of light that can illuminate a distant sign. Play around with the mirror simulation and decide how you might do this. When finished, show how the mirror equation applies to this situation. Note in particular an object distance that produces a parallel beam of light.

axis

mirror

Question 1 Thin-lens objects and images. Set the focal length of the lens to f = +60 mm (a converging lens) and with the pointer move the lens so that its middle is on a line a little to the left of the center of the simulation screen. Move the object to the left edge of the screen and then pull it slowly toward the lens. Be sure to move it inside the focal point (the blue cross on the axis of the lens). Note the changing position and size of the image.

- Which object—image positions are most like that of a camera?

- Which object—image positions are most like that of a slide projector?

- Which object—image positions are most like that of a magnifying glass?

Question 2 Thin lens ray diagrams—converging lenses. With the previous lens arrangement (f = +60 mm), set the object at the left edge of the screen and move it slowly toward the lens. Observe the two rays that represent light leaving the tip of the arrow. Develop in words a rule for how the direction of these rays is determined. Also develop a rule for determining the location of a real image on the right of the lens and a virtual image on the left.

- Path of ray 1 (on the left of lens, moves parallel to the axis of lens):

- Path of ray 2 (on the left of lens, moves toward the center of the lens):

- How do you decide where a real image is formed on the right side of the lens?

- How do you decide where a virtual image is formed on the left side of the lens?

Be sure to note that light moves away from all points on the object in all directions. The two rays in the simulation indicate the path of only a small portion of the light. The figure below shows a larger number of light rays that focus to form a real-image point.

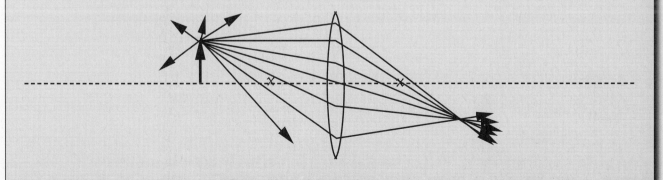

Question 3 Thin-lens ray diagrams—diverging lenses. Set the focal length to f = –60 mm to form a diverging lens. Move the object to the left edge of the screen and pull it slowly toward the lens. Observe the two rays that are shown leaving the tip of the arrow. Develop in words a rule for how the directions of these rays are determined. Also develop a rule for determining the location of the virtual image on the left side of the lens.

- Path of ray 1 (on the left of lens, moves horizontally toward the lens):

- Path of ray 2 (on the left of lens, moves toward the center of the lens):

- How do you decide where a virtual image is formed on the left side of the lens?

Question 4 Find the image. Use ray diagrams to estimate the position of the image of an object for the given object distances. Indicate the type of image (real or virtual) and the orientation of the image relative to the orientation of the object.

300-mm object distance from a +50-mm focal-length lens:

axis

mirror

80-mm object distance from a +50-mm focal-length lens:

axis

mirror

34-mm object distance from a +50-mm focal-length lens:

axis

mirror

300-mm object distance from a –100-mm focal-length lens:

axis

mirror

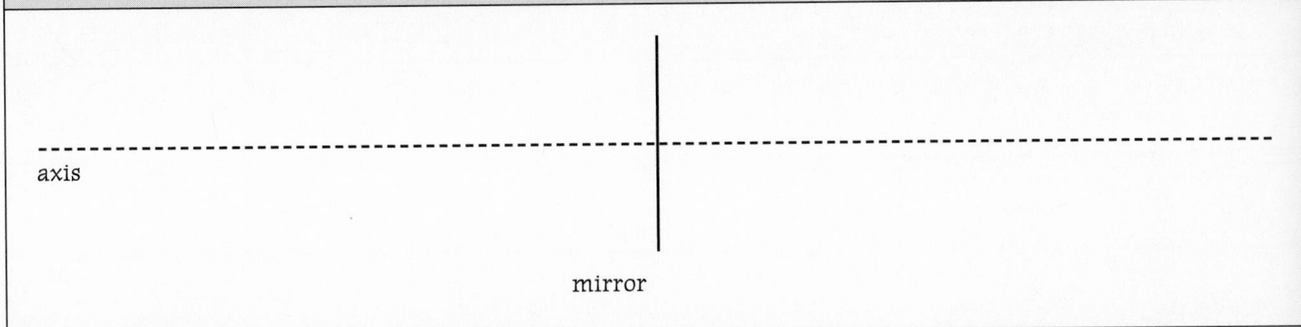

80-mm object distance from a –100-mm focal-length lens:

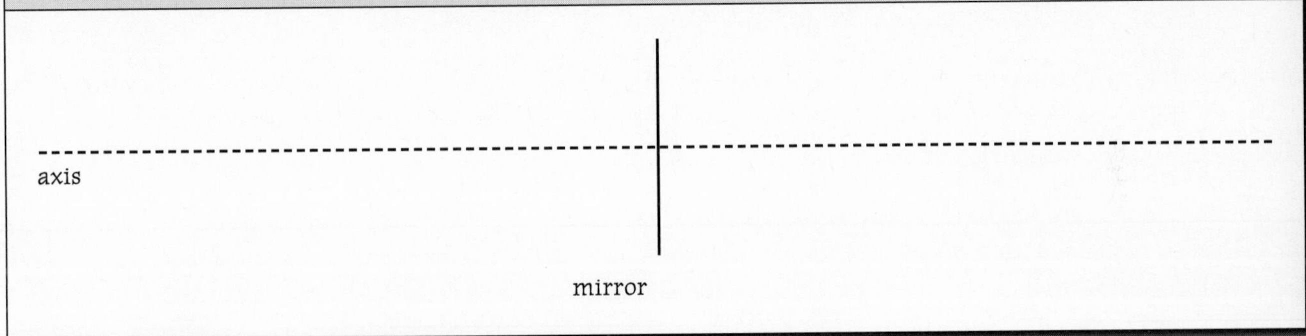

axis

mirror

Question 1 Thin lens equation—converging lens. The image distance s' is related to the focal length f of the lens and the object distance s by the following equation: $1/s + 1/s' = 1/f$. Set the focal length to f = +50 mm (a converging lens with a positive focal length). Try the following object distances and see whether the image distance is at the position predicted by the thin lens equation: (a) s = 200 mm, (b) s = 100 mm, (c) s = 60 mm, (d) s = 40 mm, and (e) s = 30 mm.

Question 2 Negative-image distances. For the last two object distances in Question 1, the image distances were negative. In the simulation, observe the image location for these negative-image distances and compare their location to those for positive-image distances. What does the negative sign mean?

Question 3 Thin-lens magnification. Use the same +50-mm focal-length lens and the same object distances as in Question 1 (200, 100, 60, 40, and 30 mm). In the simulation, note the object and image distances and the linear magnification m of the image. Invent a rule for determining the linear magnification. Interpret the meaning of the linear magnification, including its sign. If you have trouble, consider the object height and image height and orientation when s = 150 mm and when s = 30 mm.

Question 4 Camera image. A camera has a +50-mm focal-length lens. A 2.0-mm-long ant is 200 mm from the lens. Where should the film be located and how large is the image of the ant on the film?

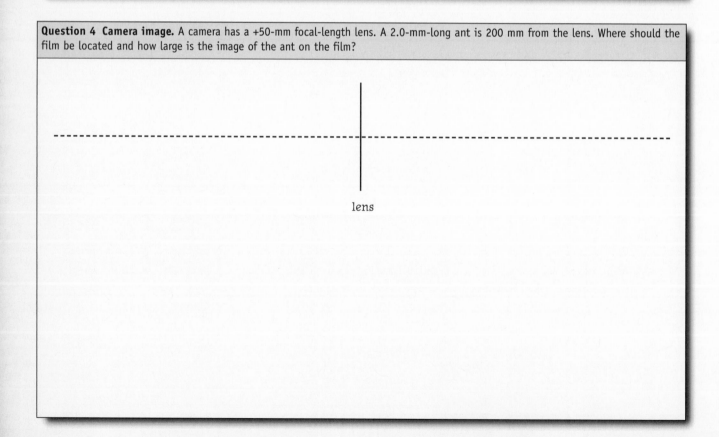

lens

Question 5 Slide projector. A slide projector has a +50-mm focal-length lens. The face of little Sarah occupies 10 mm on a slide that is inserted 55 mm from the lens. Where should the screen be located in order to form a focused image of Sarah on the screen? How large is the image of Sarah's face on the screen?

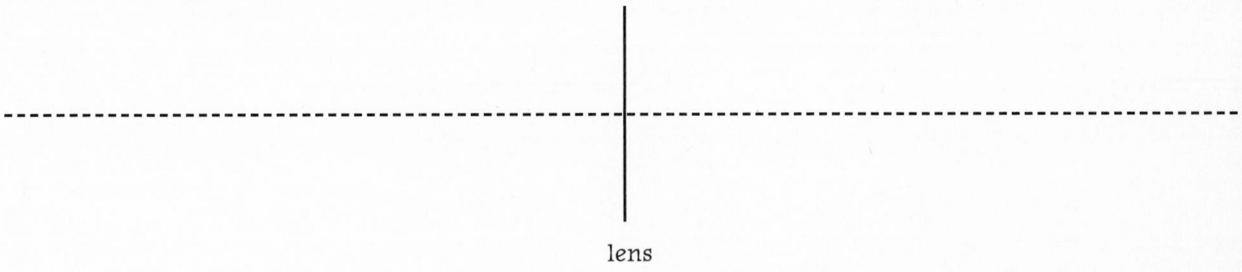

lens

Question 6 Magnifying glass. A magnifying glass has a +50-mm focal-length lens. You look through the magnifying glass at a 2.0-mm-tall ant. You would like the ant's image to be formed at the near point of your eye—200 mm to the left of the lens and from your eye. Where should the ant be located relative to the lens to produce the virtual image? How tall is the image of the ant?

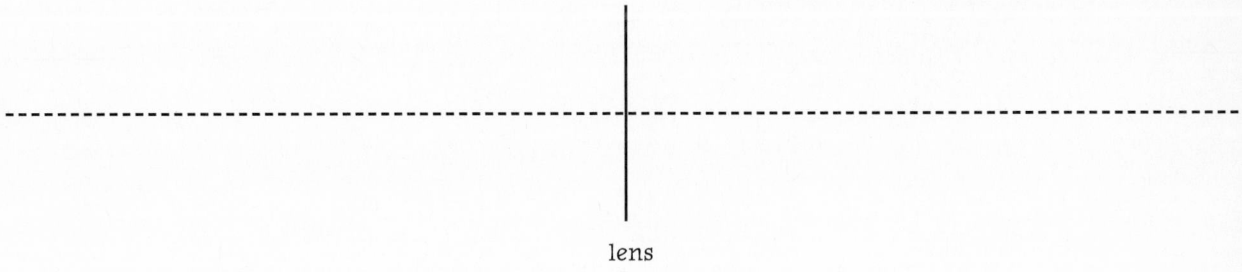

lens

Question 1 Thin-lens equation—diverging lens. The image distance s' is related to the focal length f of the lens and the object distance s by the following equation: $1/s + 1/s' = 1/f$. Set the focal length to $f = -50$ mm (a diverging lens with a negative focal length). Try the following object distances and see whether the image distance is at the position predicted by the thin-lens equation: (a) s = 200 mm, (b) s = 100 mm, (c) s = 40 mm, and (d) s = 30 mm.

Question 2 Negative-image distance. The image distances are negative for all of the object locations in Question 1. Note the image location for these negative-image distances. What does the negative sign mean?

Question 3 Thin-lens magnification. Use the same –50-mm focal-length lens and two of the object distances from Question 1 (200 mm and 40 mm). In the simulation, note the object and image distances and the linear magnification m of the image. Qualitatively, do these image and object heights seem to be related by the linear magnification equation used in Activity 15.10? Explain.

$$m = \frac{\text{image height}}{\text{object height}} = \frac{h'}{h} = -\left(\frac{\text{image distance}}{\text{object distance}}\right) = -\left(\frac{s'}{s}\right)$$

Question 4 Nearsighted glasses. Your nearsighted pet mouse can focus on objects that are no farther away than 12 cm (120 mm). The mouse needs to see its cheese, which is 30 cm away. What focal-length lenses for the mouse eyeglasses will form an image 12 cm from the glasses for the 30-cm-distant cheese? If the cheese is 10 mm tall, how tall is the image of the cheese as seen through the lenses?

lens

Question 1 Two-lens image location. Set the focal length of lens 1 to +40 mm (+0.040 m) and the focal length of lens 2 to +35 mm (+0.035 m). Set x_1 = 75 mm and x_2 = 275 mm. We now use a general procedure to find the final image produced by the two-lens system.

First image location: Use the thin-lens equation to find the image produced by the first lens. The object of the first lens is 75 mm left of the +40-mm focal-length lens.

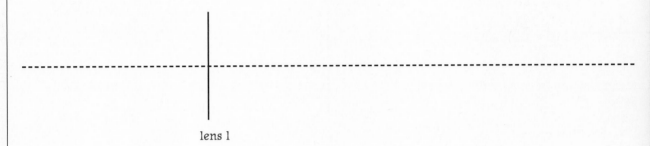

lens 1

Second object location: The image of the first lens is the object of the second lens. To find this object distance, we must determine the distance of the first image from the second lens. In this example, the lenses are separated by $x_2 - x_1$ = 275 mm – 75 mm = 200 mm. Since the image of the first lens is a distance s_1' = 86 mm to the right of the first lens, its distance from the second lens is 200 mm – 86 mm = 114 mm. In general, the object distance for the second lens is $s_2 = (x_2 - x_1) - s_1'$.

Second image location: Use the thin-lens equation to find the image produced by the second lens. The object of the second lens is 114 mm left of the 35-mm focal-length lens.

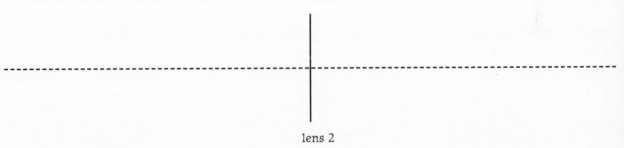

lens 2

Question 2 Image magnification. Determine the linear magnification for the two-lens system.

$$m_{net} = \frac{\text{final image height}}{\text{original object height}} = \frac{h_2'}{h_1} = -\left(\frac{s_1'}{s_1}\right)\left(\frac{s_2'}{s_2}\right)$$

Question 3 Locating and sizing another two-lens image. Set the focal lengths for both lenses to +50 mm. Set $x_1 = 60$ mm and $x_2 = 200$ mm.

First image location: Use the thin-lens equation to find the image produced by the first lens. The object of the first lens is 60 mm left of the +50-mm focal-length lens.

lens l

Second object location: The image of the first lens is the object of the second lens. Determine the object distance for the second lens. **Note:** If you get a negative sign for s_2, it means the object is on the right side of the lens—on the right side after the light has passed through the lens. Strange! Include the negative sign in your next calculation. In general, the object distance for the second lens is: $s_2 = (x_2 - x_1) - s_1'$.

Second image location: Use the thin-lens equation to find the image produced by the second lens.

Image Magnification: Determine the linear magnification for the two-lens system.

$$m_{net} = \frac{\text{final image height}}{\text{original object height}} = \frac{h_2'}{h_1} = -\left(\frac{s_1'}{s_1}\right)\left(\frac{s_2'}{s_2}\right)$$

Question 4 Microscope. Set the focal length of the first lens to +50 mm and the focal length of the second lens to −100 mm. Set the first lens at x_1 = 60 mm and the second lens at x_2 = 200 mm. Determine the position of the final image and the magnification of the two-lens system.

First image location: Use the thin-lens equation to find the image produced by the first lens.

lens 1

Second object location: The image of the first lens is the object of the second lens. Determine the object distance for the second lens. In general, the object distance for the second lens is: $s_2 = (x_2 - x_1) - s_1'$.

Second image location: Use the thin-lens equation to find the image produced by the second lens.

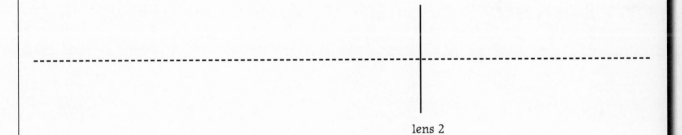

lens 2

Image Magnification: Determine the linear magnification for the two-lens system.

$$m_{net} = \frac{\text{final image height}}{\text{original object height}} = \frac{h_2'}{h_1} = -\left(\frac{s_1'}{s_1}\right)\left(\frac{s_2'}{s_2}\right)$$

Question 1 The telescope. Set the focal length of the first objective lens to +150 mm and the focal length of the second eyepiece lens to + 30 mm. The distance between the lenses is adjusted automatically to the sum of the focal lengths (150 mm + 30 mm = 180 mm in this example). Adjust the angle of the light from the distant object to +0.05 rad. We now use our general two-lens image location procedure to find the final image of that distant object.

First image location: Use the thin-lens equation to find the image produced by the first lens.

Second object location: The image of the first lens is the object of the second lens. Determine the object distance for the second lens.

Second image location: Use the thin-lens equation to find the image produced by the second lens.

Question 2 Angular magnification of a telescope. Suppose the object of the first lens is off axis so that incoming rays make an angle θ_1 with the axis. The ray that passes through the middle of a thin lens does not change direction and passes through the first image point.

From the drawing, we see that

(1) $\tan \theta_1 = h_1'/s_1' = h_1'/f_1$

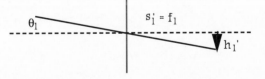

since $s_1' = f_1$. The image of lens 1 is the object for lens 2. Thus the height of object 2, h_2, equals the height of image 1 h_1'. Light leaving object 2 is shown in the sketch below. The angle θ_2 that it makes with respect to the optical axis is determined from the following equation:

(2) $\tan \theta_2 = h_2/s_2 = h_2/f_2$

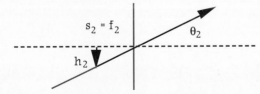

since $s_2 = f_2$. Combining Eqs. (1) and (2) and using the small angle approximation ($\tan \theta \approx \theta$ for small angles), we find that $\theta_1 \approx h_2/f_1$ and $\theta_2 \approx h_2/f_2$. The **angular magnification of the telescope** is $M = -\theta_2/\theta_1 = -f_1/f_2$. For this problem, the angular magnification is −5.0.

Question 1 Accommodation.

Set the object distance slider to s = 30 cm. Press on the "Normal" eye button. Leave the other buttons off (unchecked). You will see two light rays representing light coming from an object that is 30 cm in front of the eye. The light passes through the lens of the eye and is focused as a real image on the retina at the back of the eye.

Now, press the "Infinity" button on and off. The object will jump between infinity and 30 cm. As you do this, watch the image location, then the shape of the lens of the eye, and finally the focal length of the lens. This is called **accommodation.** Does the image position change? If not, how does light from an object 30 cm in front of the lens and light from an object infinitely far from the lens manage to focus at the same place?

Question 2 Focused and unfocused images.

Turn on the "Nearsighted" eye button. Place the object slider at 12 cm (s = 12 cm). The lens of the eye forms a pointlike blue image of the object at the surface of the back of the eye—at the retina. Now, leave the object at the same position and turn on the "Farsighted" eye button. The image is now formed somewhat behind the retina. Why would this image appear blurred or unfocused to the eye?

Question 3 Near point.

Move the object distance slider to s = 40 cm. Turn on the "Normal" type of eye button. Now, move the object distance slider (the s slider) so that the object moves closer to the lens of the eye. What is the nearest object distance **(the near-point distance)** for which the image is still focused on the retina? As the object moves closer to the lens than this near point, does the image move farther in front of the retina or farther behind it?

Question 4 Near point for other types of vision.

(a) Find the near-point distance for the farsighted person in the simulation.

(b) Find the near-point distance for the nearsighted person in the simulation.

(c) Estimate the near-point distance for your own eye.

Question 5 Far point.

(a) Move the object distance slider to s = 50 cm. Leave the "Glasses off" and turn on the "Nearsighted" type of eye button. Now, slowly move the object distance slider so that the object moves farther from the lens of the eye. What is the farthest object distance (the **far-point distance**) for which the image is still focused on the retina? If the object is farther from the lens than the far point, where does the image form—in front of the retina or behind it?

(b) Find the far-point distance for the simulation's normal eye.

(c) Find the far-point distance for the simulation's farsighted eye.

Question 6 Thin-lens equation for eye's vision. Create and solve two different problems that show that the simulation is consistent with the thin-lens equation. Give all of the numbers and settings used and show how they are consistent with the equation.

Question 1 Glasses for a farsighted person to see near objects.

Make the following simulation settings: "Farsighted" eye, "Glasses off" and "s = 20 cm" object distance. Light from the nearby object passes through the lens of the *eye* and forms an unfocused image *behind* the retina. What type of eyeglass lens allows the *eye* to form a focused image on the retina? You can answer this question in a special way by answering the following questions.

(a) Determine the near-point distance of this farsighted eye.

(b) Use the thin-lens equation to determine the focal length of eyeglass lens that forms an image of the near object (s = 20 cm) to the left of the eyeglass lens and at (or beyond) the eye's near point. This distance image formed by the eyeglass lens becomes the object of the eye's lens. This new object is far from the eye, and the farsighted eye can produce a focused image of this new object that is formed by the eyeglasses.

Question 2 Glasses for a nearsighted person to see distant objects.

Make the following simulation settings: "Nearsighted" type of eye, "Glasses off," and "Infinity" checked. Light from the distant object passes through the lens of the *eye* and forms an unfocused image in *front of* the retina. What type of eyeglass lens allows the *eye* to form a focused image on the retina? You can answer this question in a special way by answering the following questions.

(a) Determine the far-point distance of this nearsighted eye.

(b) Use the thin-lens equation to determine the focal length of eyeglass lens that forms an image of a distant object (s = infinity) to the left of the eyeglass lens and at (or nearer than) the eye's far point. This nearby image of the eyeglass lens becomes the object of the eye's lens. Since the new object is nearer the eye, the nearsighted eye can produce a focused image of this nearer object.

16

PHYSICAL OPTICS

Question 1 The central n = 0 bright band. Set the wavelength to 400 nm and the source separation to d = 2.0 mm. The light from the two sources when it reaches the screen causes alternating bright bands and dark bands. Why is the light bright at position y = 0.0, the center of the screen? This is called the n = 0 bright band.

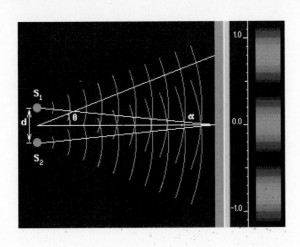

Question 2 The n = 1 bright band. With λ = 400 nm and d = 2.0 mm, use the figure above to help explain why the light is again bright at position y = 0.80 mm.

Question 3 The n = 2 bright band. With λ = 400 nm and d = 2.0 mm, explain why the light would again be bright at position y = 1.60 mm (not shown in the figure above or in the simulation).

Question 4 The angular deflection θ_n of the nth bright band. Use the figure below to see that the angular deflection θ_n of the light traveling toward the nth bright band is given by the equation $\sin \theta_n = n\lambda/d$, where d is the separation of the light sources and λ is the wavelength of the light.

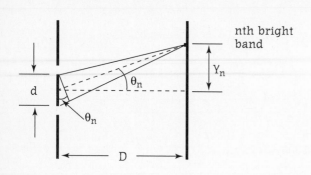

Question 5 Bright-band positions on a screen. Use the figure above to see that the position y_n of the nth bright band on the screen is given by the equation $\tan \theta_n = y_n/D$, where D is the distance of the light sources from the screen and y_n is the distance on the screen of the nth bright band from the central maximum.

Question 6 Small-angle approximation. You just found that $\sin \theta_n = n\lambda/d$ and $\tan \theta_n = y_n/D$. For small angles (10° or less), $\sin \theta = \tan \theta$. This approximation is usually true for two-source interference involving light but is not true in general for longer wavelength two-source interference (e.g., for sound waves). Assuming this is true for two-source interference of light waves, show that the separation of fringes on the screen is $\Delta y = \lambda D/d$.

Question 7 Predict the separation of the bright bands. You just found that the bright bands are separated by a distance $\Delta y = \lambda D/d$. Predict the separation of the bright bands for two 500-nm turquoise light sources separated by 2.0 mm when observed on a screen 4.0 m from the sources.

Question 1 Effect of source separation on the band pattern. Set the wavelength to $\lambda = 600$ nm and the source separation to d = 3.0 mm. The light from the two sources when it reaches the screen causes alternating bright bands and dark bands. The pattern is shown on the right side of the simulation screen.

When the sources are moved closer to each other, the position y_1 of the first bright band from the central maximum at $y_0 = 0.0$ **(a)** increases, **(b)** remains the same, or **(c)** decreases. Justify your answer.

Question 2 Effect of wavelength on the band pattern. Set the wavelength to $\lambda = 400$ nm and the source separation to d = 2.0 mm. The light from the two sources when it reaches the screen causes alternating bright bands and dark bands. The pattern is shown on the right side of the simulation screen. Note the separation Δy of the central bright band and the first bright band above it.

When the wavelength of the light is increased, the position y_1 of the first bright band from the central maximum **(a)** increases, **(b)** remains the same, or **(c)** decreases. Justify your answer.

Question 1 Separation of bands on a screen. Set the wavelength to $\lambda = 400$ nm (violet light) and the source separation to $d = 2.0$ mm. The light from the two sources when it reaches the screen 4.0 m from the sources causes alternating bright bands and dark bands. The distance Δy between the central bright band and the first bright band at its side (above it on the simulation screen) is 0.80 mm. Determine the distance between the central bright band and the first bright band if the wavelength is changed to 630 nm.

Question 2 Separation of bands on a screen. Determine the distance on the screen between the central bright band and the first bright band with the light wavelength at 630 nm and the slit separation now adjusted to 3.0 mm.

Question 3 Separation of bands on a screen. By what fraction will the separation of the central bright band and the first bright band change if the wavelength is decreased from 630 nm to 420 nm and the separation of the sources is decreased from 3.0 mm to 1.0 mm? Start by recording the initial value of y_1. Then, try to calculate in your head the change in y_1 when the wavelength and source separation are made. Finally, check the final y_1 by changing the simulation sliders.

Question 4 Separation of bands on a screen. You are asked to help with an art exhibit in which you are to provide on a screen bands of 500-nm turquoise light that are separated by 2.0 mm. Determine the separation between two synchronized light sources that produces the desired bands on a screen 4.0 m from the sources.

Question 5 Separation of bands on a screen. You find that you can only produce bands on a screen for the art exhibit from sources separated by 3.0 mm. What wavelength light should you use to provide bands on the screen with a separation of 0.92 mm for a screen 4.0 m from the sources?

Question 6 Angular deflection and position on screen of nth bright band. (a) Determine the angular deflection to the fourth bright band to the side of the central maximum for 600-nm yellow light sources separated by 2.0 mm. **(b)** Determine the distance from this fourth bright band to the side from the central maximum when observed on a screen 4.0 m from the source.

Question 1 The central m = 0 bright band. Set the wavelength to 600 nm (yellow light) and the source separation to $d = 2.0 \times 10^{-3}$ mm. Why do we see bright light straight ahead at the center of the screen (at the 0.0 position)? This is called the m = 0 bright band. Does the position of this band depend on the wavelength of the light or on the slit separation? Explain.

Question 2 The m = 1 bright band. With $\lambda = 600$ nm and $d = 2.0 \times 10^{-3}$ cm, why is the light bright at position $y_1 = 1.50$ cm?

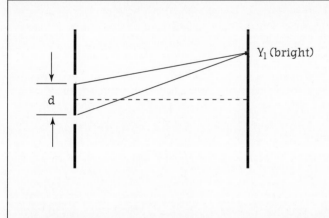

Y_1 (bright)

d

Question 3 The m = 2 bright band. With $\lambda = 600$ nm and $d = 2.0 \times 10^{-3}$ cm, why is the light bright at position $y_2 = 3.0$ cm?

Question 4 The angular deflection θ_m of the nth double-source bright band. Explain why light traveling toward the mth bright band at an angular deflection θ_m is given by the equation $\sin \theta_m = m \lambda/d$, where d is the separation of the light sources and λ is the wavelength of the light. (**Hint:** Look at the figure in Question 4 in the worksheet for Activity 16.1.)

Question 5 Bright-band positions on a screen. Explain why light traveling from the grating toward the mth bright band on the screen is given by the equation $\tan \theta_m = y_m/D$, where D is the distance of the grating from the screen (in the simulation, D = f where f is the focal length of the lens) and y_m is the distance on the screen of the mth bright band from the central maximum.

Question 6 Small-angle approximation. For small angles (10° or less), $\sin \theta = \tan \theta$. This approximation applies for the simulation geometry (note that y_m is much less than $D = f$ and θ_m is small). However, for many real gratings, the angular deflection of the first and second bright bands may be large and the two equations must be used separately. Assuming that the angular deflection is small, show that the separation of bright bands on the screen is $\Delta y = \lambda f/d$, where the distance D to the screen equals the lens focal length f.

Question 7 Effect of source separation on the band pattern. Set the wavelength to 600 nm (yellow light) and the source separation to $d = 3.0 \times 10^{-3}$ cm. The light from the the grating slits, when it reaches the screen, causes a series of bright bands. The pattern is shown on the right side of the simulation screen.

When the slits are moved closer to each other, the position y_1 of the first bright band from the central maximum at $y_o = 0.0$ **(a)** increases, **(b)** does not change, or **(c)** decreases. Why does this happen?

Question 8 Effect of wavelength on the band pattern. Set the wavelength to 400 nm (violet light) and the source separation to d = 2.0×10^{-3} cm. The light from the slits when it reaches the screen causes several bright bands. The pattern is shown on the right side of the simulation screen. Note the separation y_1 of the central bright band and the first bright band above it.

When the wavelength of the light is increased, the position y_1 of the first bright band from the central maximum **(a)** increases, **(b)** does not change, or **(c)** decreases. Why does this happen?

Question 1 Separation of bands on a screen. Set the wavelength to 420 nm (violet light) and the source separation to $d = 2.1 \times 10^{-3}$ cm. The light passing through the slits reaches the screen 50 cm from the slits and causes on the screen a series of bright bands separated from each other by 1.0 cm. **(a)** Show that this band separation is predicted by one of the grating equations. **(b)** Determine the distance on the screen between the central bright band and the first bright band at the side if the wavelength is increased from 420 nm to 630 nm, the wavelength of red light from a helium–neon laser.

Question 2 Separation of bands on a screen. With the wavelength still at 630 nm, if you change the slit separation from 2.1×10^{-3} cm to 1.4×10^{-3} cm (don't do it yet), what wavelength of light causes the distance on the screen between the central bright band and the first bright band at the side to remain at 1.5 cm?

Question 3 Separation of bands on a screen. Determine the separation of the central bright band and the first bright band at the side if the wavelength is 690 nm and the separation of the slits is 3.0×10^{-3} cm. When finished, check the value of y_1 by changing the simulation sliders.

Question 4 Separation of bands on a screen. You are asked to help with an art exhibit in which you are to provide on the screen bright bands of 480-nm turquoise light with a separation of 2.0 cm between adjacent bands. Determine the separation between grating slits that produces the desired bands on a screen 50 cm from the slits. (The light passes through a 50-cm focal-length lens just after leaving the grating.) When finished, adjust the sliders to check your work.

Question 5 Separation of bands on a screen. You find that you can only produce bands on a screen for the art exhibit from sources separated by 2.5×10^{-3} cm. What wavelength light should you use to provide bands with a separation of 1.2 cm on a screen 50 cm from the sources? When finished, adjust the sliders to check your work.

Question 6 Angular deflection and position on screen of fourth bright band. (a) Determine the angular deflection of the fourth bright band to the side of the central maximum for 600-nm yellow light passing through a grating with 400 slits/cm. **(b)** Determine the distance of this fourth bright band to the side of the central maximum when observed on a screen 50 cm from the source. When finished, set the simulation to these settings to check your answers.

Question 1 Effect of wavelength on the first dark band m = 1 position on the screen. Set the wavelength to 400 nm (violet light) and the slit width to d = 0.30 mm. Note the positions of the first m = 1 dark bands on each side of the central maximum. Design your own experiments to determine how the positions of these m = 1 dark bands are affected by the wavelength of light. Provide numbers that support your conclusion.

Question 2 Effect of slit width on the first dark band m = 1 positions on the screen. Set the wavelength to 500 nm (turquoise light) and the slit width to d = 0.50 mm. Note the positions of the m = 1 dark bands on each side of the central maximum. Design your own experiments to determine how the positions of these m = 1 dark bands are affected by the slit width d. Provide numbers that support your conclusion.

SUMMARY: **The Mathematics of Single-Slit Diffraction**

- **Angular deflection to bright bands.** The angular deflection θ_m from the center of the slit to the mth dark band on the screen is determined from the equation $\sin \theta_m = m \lambda/d$, where λ is the wavelength of the light and d is the width of the slit.

- **Dark-band position on screen.** The position x_m on a screen of the center of the mth dark band from the center of the central maximum is determined from the equation $\tan \theta_m = x_m/L$, where θ_m is the angular deflection from the center of the slit to the mth dark band and L is the distance from the slit to the screen.

- **Small-angle approximation.** Often, the angular deflection of light to the dark bands is very small. In that case, $\sin \theta_m = \tan \theta_m$. We can set the two expressions above equal to each other and obtain the equation $m \lambda/d = x_m/L$. This equation can be used to quickly determine one of the quantities if the other quantities are known.

Question 3 Dark-band positions on a screen. Predict the location of the first and second dark bands on the screen and the angular deflection from the slit to these dark bands for 600-nm light passing through a 0.40-mm slit. The screen is 10 m from the slits. After your prediction, move the sliders to check your prediction.

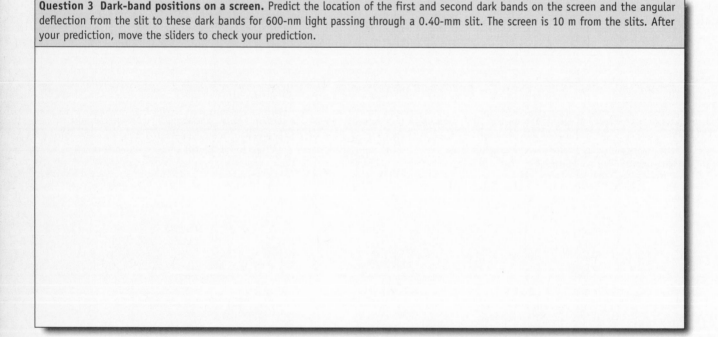

©1999 Addison Wesley Longman, Alan Van Heuvelen and Paul D'Alessandris

Question 4 Wavelength. Predict the wavelength of light that produces the third dark band (the m = 3 dark band) 30 mm from the center of the central bright band after passing through a 0.40-mm wide slit. The screen is 10 m from the slits. After your prediction, move the sliders to check your prediction.

Question 5 Wavelengths and slit widths. Use the diffraction equations to predict three different wavelength and slit width combinations that produce the second dark band 30 mm from the center of the central bright band. The screen is 10 m from the slits. After your predictions, move the sliders to check your predictions.

Question 1 The m = 1 dark ring on the screen. Set the wavelength to 700 nm (red light) and the hole radius to R = 0.30 mm. Note the radius of the first m = 1 dark ring on the screen. Design your own experiment to determine how the radius of the m = 1 dark ring on the screen is affected by the wavelength of light. Provide approximate numbers to support your conclusion.

Question 2 The m = 1 dark ring on the screen. Set the wavelength to 600 nm (yellow light) and the hole radius to R = 0.30 mm. Note the radius of the m = 1 dark ring on the screen. Design your own experiment to determine how the radius of the m = 1 dark ring on the screen is affected by the radius R of the hole. Provide approximate numbers that support your conclusion.

SUMMARY: **The Mathematics of Circular-Hole Diffraction**

- **Angular deflection to dark rings.** The angular deflection θ_m from the center of the hole to the mth dark ring on the screen is determined from the equation $\sin \theta_m = 1.22\, m\, \lambda/2R$, where λ is the wavelength of the light and R is the radius of the circular hole.

- **Dark band position on screen.** The radius r_m on a screen of the mth dark ring is determined from the equation $\tan \theta_m = r_m/L$, where θ_m is the angular deflection from the center of the circular hole to the mth dark ring and L is the distance from the hole to the screen.

- **Small-angle approximation.** Often, the angular deflection of light to the dark rings is very small. In that case, $\sin \theta_m = \tan \theta_m$. We can set the two expressions above equal to each other and obtain the equation $1.22\, m\, \lambda/2R = r_m/L$. This equation can be used to quickly determine one of the quantities if the other quantities are known.

Question 3 Dark ring positions on a screen. Predict the location of the first dark ring on the screen and the angular deflection from the hole to this dark ring for 500-nm turquoise light passing through a 0.19-mm hole. The screen is 10 m from the slits. After your prediction, move the sliders to check your prediction.

©1999 Addison Wesley Longman, Alan Van Heuvelen and Paul D'Alessandris

Question 4 Radius of hole. When 630-nm red light shines on a small circular hole, a diffraction pattern is formed with a first dark-ring radius of 25.6 mm (about two inches in diameter). The screen is 10 m from the slits. Determine the radius of the circular hole. After your prediction, move the sliders to check your prediction. **Note:** The dimensions of very small objects can be measured by diffraction. For example, scientists at Los Alamos Scientific Labs measured the radius of body cells and the radius of the nuclei of these cells using diffraction. This diffraction was used to develop a technique to detect cervical cancer.

Question 5 Wavelengths and circular-hole radii. Use the diffraction equations to predict three different wavelength and circular-hole radius combinations that produce an m = 1 dark ring of radius 21 mm. The screen is 10 m from the slits. After your predictions, move the sliders to check your predictions.

Question 1 Effect of angular separation of sources on resolving power. Start with the wavelength at $\lambda = 600$ nm, the shutter opening at $D = 1.5$ cm, and the angular separation of the sources $\theta = 10.0 \times 10^{-5}$ rad. Can you see on the screen the image of the two point objects? Now, decrease only the angular separation θ of the sources in several steps to the minimum allowed value. Observe the pattern formed on the screen. How does decreasing the angular separation of the sources affect your ability to "see" the images of the two objects on the screen? Explain.

Question 2 Effect of shutter size on resolving power. Set the wavelength to $\lambda = 600$ nm, the shutter opening to $D = 1.5$ cm, and the angular separation of the sources to $\theta = 5.0 \times 10^{-5}$ rad. Can you see the images of the two point objects on the screen? How does decreasing the shutter size to 1.0 cm and 0.50 cm affect your ability to "see" the images of the two objects on the screen? Explain.

Question 3 Effect of wavelength on resolving power. Set the wavelength to $\lambda = 400$ nm, the shutter opening to $D = 0.60$ cm, and the angular separation of the sources to $\theta = 10.0 \times 10^{-5}$ rad. Can you see the image of the two point objects on the screen? Now, increase only the wavelength λ in 100-nm steps and observe the pattern formed on the screen. How does increasing the wavelength affect your ability to resolve the images of the two objects on the screen? Explain.

Question 4 The separation of the images on a screen. Set the wavelength to 600 nm (yellow light), the shutter opening to D = 1.4 cm, and the angular separation of the sources to 5.0×10^{-5} rad. Decide whether the centers of the images on the screen are the correct distance apart. The screen is 20 cm from the lens.

Question 5 Width of the images. The 600-nm light from the point sources illuminates the circular opening with shutter diameter d = 1.4 cm. Images are formed on a screen 20 cm from the lens. Use the circular-hole diffraction equation to determine the radius of the first dark ring surrounding one of the images on the screen. Then, compare your work with the image size of one image on the simulation screen.

The **resolving power** of an optical instrument is a measure of its ability to produce separate images of two nearby point objects. Diffraction is the ultimate limit of resolving power. The accepted criterion for resolution is the Raleigh criterion:

$$\theta_{\text{angular separation of objects and images}} > 1.22 \, \lambda/D$$

According to this criterion, two images are just resolved if the center of the central maximum of one diffraction pattern falls on the first dark ring of the other diffraction pattern.

Question 6 Binary stars. Predict the minimum shutter diameter that can resolve the 400-nm violet light from binary stars with an angular separation of 3.6×10^{-5} rad. When finished with your prediction, set the shutter diameter to this value and see whether the images are barely observable.

Question 7 Moon rocks. Your home telescope has a maximum shutter opening of 1.5 cm. Determine the closest distance of two shiny rocks on the moon 3.8×10^8 m from the earth that can be resolved with your telescope. Assume that the telescope is most sensitive to 600-nm yellow light. When finished with your prediction, set the sliders to check your work.

Question 8 Car headlights. Estimate the maximum distance that two car headlights can be for you to distinguish the lights while looking through a 1.0-cm-diameter telescope lens. Assume that the light wavelength is 550 nm. When finished with your estimate, set the slider settings to check your work.

Question 1 Angle dependence of transmitted intensity. In the wave theory of light, the light intensity I is proportional to the square of the amplitude A of the light wave I = (constant) A^2. Set the orientations of both Polaroids to 0°. The difference in the orientation of the two Polaroids is given the symbol $\Delta\phi$. Observe the changing intensity of the light leaving the second Polaroid as you turn the second Polaroid from 0° to 90°. Confirm that the intensity of the light decreases by $\cos^2 \Delta\phi$ and not by $\cos \Delta\phi$.

Question 2 Transmitted intensity for different Polaroid orientations. Unpolarized light of intensity I_o is incident on the first of two Polaroid films. **(a)** If you set the Polaroid orientations to $\phi_1 = \phi_2 = 16°$, what is the intensity of the transimitted light? After your prediction, check your results with the simulation.

(b) If you set the Polaroid orientations to $\phi_1 = 16°$ and $\phi_2 = 32°$, what is the intensity of the transmitted light? After your prediction, check your results with the simulation.

(c) If you set the Polaroid orientations to $\phi_1 = 16°$ and $\phi_2 = 46°$, what is the intensity of the transmitted light? After your prediction, check your results with the simulation.

(d) If you set the Polaroid orientations to $\phi_1 = 16°$ and $\phi_2 = 76°$, what is the intensity of the transmitted light? After your prediction, check your results with the simulation.

(e) If you set the Polaroid orientations to $\phi_1 = 16°$ and $\phi_2 = 106°$, what is the intensity of the transmitted light? After your prediction, check your results with the simulation.

Question 3 Angle for certain transmitted intensity. Unpolarized light of intensity I_0 is incident on the first of two Polaroid films. **(a)** Set the orientation for the first Polaroid to $\phi_1 = 30°$. What orientation should the second Polaroid have so that the transmitted intensity through both Polaroids is 0.5 I_0? After your prediction, check your results with the simulation.

(b) Set the orientation for the first Polaroid to $\phi_1 = 30°$. What orientation should the second Polaroid have so that the transmitted intensity through both Polaroids is 0.3 I_0? After your prediction, check your results with the simulation.

(c) Set the orientation for the first Polaroid to $\phi_1 = 30°$. What orientation should the second Polaroid have so that the transmitted intensity through both Polaroids is 0.1 I_0? After your prediction, check your results with the simulation.

Question 4 Transmitted intensity for different Polaroid orientations. Unpolarized light of intensity I_0 is incident on the first of two Polaroid films. **(a)** If you set the Polaroid orientations to $\phi_1 = 30°$ and $\phi_2 = 76°$, what is the intensity of the transmitted light? After your prediction, check your results with the simulation.

(b) With the second Polaroid orientation still at $\phi_2 = 76°$, what is the transmitted intensity through the second Polaroid if you now change the first Polaroid's orientation from $\phi_1 = 30°$ to $\phi_1 = 16°$? After your prediction, check your results with the simulation.

(c) With the first Polaroid orientation still at $\phi_1 = 16°$, what is the transmitted intensity through the second Polaroid if you now change its orientation from $\phi_2 = 76°$ to $\phi_2 = 136°$? After your prediction, check your results with the simulation.

©1999 Addison Wesley Longman, Alan Van Heuvelen and Paul D'Alessandris

MODERN PHYSICS

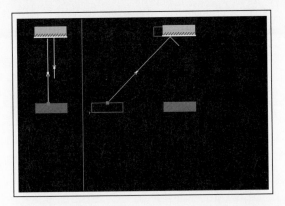

Question 1 Distance traveled by the light pulse, as measured on the earth. How does the distance traveled by the light pulse on the moving light clock compare to the distance traveled by the light pulse on the stationary light clock?

Question 2 Time interval required for light pulse travel, as measured on earth. Given that the speed of the light pulse is independent of the speed of the light clock, how does the time interval for the light pulse to travel to the top mirror and back on the moving light clock compare to on the stationary light clock?

Question 3 Time interval required for light pulse travel, as measured on the light clock. Imagine yourself riding on the light clock. In your frame of reference, does the light pulse travel a larger distance when the clock is moving, and hence require a larger time interval to complete a single round-trip?

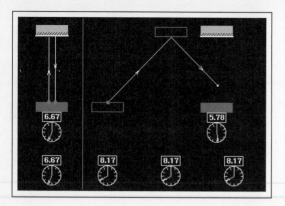

Question 4 The effect of velocity on time dilation. Will the *difference* in light pulse travel time between the earth's timers and the light clock's timers increase, decrease, or stay the same as the velocity of the light clock is decreased?

Will the distance the light pulse travels, as measured by an observer riding on the light clock, change?	Will the distance the light pulse travels, as measured by an observer on the earth, change?

Will the time of travel, as measured by the light clock's timers, change?	Will the time of travel, as measured by the earth's timers, change?

Final answer:

Question 5 The time dilation formula. Using the time dilation formula, predict how long it will take for the light pulse to travel to and fro between mirrors, as measured by an earth-bound observer, when the light clock has a Lorentz factor (γ) of 1.2. The proper time (Δt_{proper}) is

The Lorentz factor (γ) is

The time measured on the earth (Δt) is

Question 6 The time dilation formula, one more time. If the time interval between departure and return of the light pulse is measured to be 7.45 ms by an earth-bound observer, what is the Lorentz factor of the light clock as it moves relative to the earth? The proper time (Δt_{proper}) is

The time measured on the earth (Δt) is

The Lorentz factor (γ)

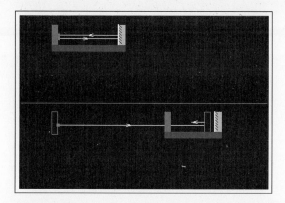

Question 1 Round-trip time interval, as measured on the light clock. Imagine riding on the left end of the light clock. A pulse of light departs the left end, travels to the right end, reflects, and returns to the left end of the light clock. Does your measurement of this round-trip time interval depend on whether the light clock is moving or stationary relative to Earth? Explain

Question 2 Round-trip time interval, as measured on the earth. Will the round-trip time interval for the light pulse as measured on the earth be longer, shorter, or the same as the time interval measured on the light clock?

Which time interval, the one measured by the light clock's timer or the earth's timers, is the proper time interval?

Question 3 Why does the moving light clock shrink? You have probably noticed that the length of the moving light clock is smaller than the length of the stationary light clock. Is the round-trip time interval measured on the earth currently equal to the product of the proper time interval and the Lorentz factor?

Could the round-trip time interval as measured on the earth be equal to the product of the Lorentz factor and the proper time interval if the moving light clock were the same size as the stationary light clock?

Question 4 The length contraction formula. A light clock is 1000 m long when measured at rest. How long would an earth-bound observer measure the clock to be if it had a Lorentz factor of 1.3 relative to the earth? The proper length (l_{proper}) is

The Lorentz factor (γ) is

The length measured by an earth-bound observer (l) is

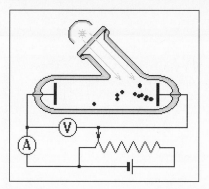

Question 1 The effect of intensity. What will happen to the number of electrons that are ejected from the metal as the intensity (brightness) of the light is increased? Explain.

Question 2 Positive applied potential. What will happen to the current if the battery provides a positive potential on the collecting plate relative to the emitting plate? Explain.

Question 3 Negative applied potential. What will happen to the current if the battery provides a negative potential on the collecting plate relative to the emitting plate? Explain.

Question 4 Energy conservation. Consider the motion of an electron from when it first leaves the emitting plate until it is stopped and turned around by the action of the stopping potential. The electron just barely misses making it to the collecting plate.

Construct an energy bar chart for this motion, defining the emitting plate as zero potential.

Initial Energy + Work =   Final Energy

$$K \; + \; U_g \; + \; U_E \; + \; W$$

$$K \; + \; U_g \; + \; U_E \; + \; U_{in}$$

0   0

Translate your bar chart into a mathematical relationship between the electron's kinetic energy when it leaves the emitting plate and the applied potential difference.

Question 5 Stopping potential vs. intensity. If you increase the intensity of the incoming light, will the stopping potential *increase, decrease,* or *stay the same?* If you increase the intensity, what happens to the number of photons?

If you increase the intensity, what happens to the energy of each photon?

Does the stopping potential depend on the number or energy of the photons?

Final answer:

Question 6 Intensity vs. current. If you decrease the intensity of the incoming light, will the current *increase, decrease, or stay the same?* If you decrease the intensity, what happens to the number of photons?

If you decrease the intensity, what happens to the energy of each photon?

Does the current depend on the number or energy of the photons?

Final answer:

Question 7 Stopping potential vs. wavelength. If you decrease the wavelength of the incoming light, will the stopping potential *increase, decrease* or *stay the same?* As the wavelength is decreased, what happens to the photon energy?

As the wavelength is decreased, what happens to the electron energy (after absorption)?

As the wavelength is decreased, what happens to the potential needed to stop the electron?

Question 8 Threshold wavelength.

What happens to the strength with which electrons are bound to the metal as wavelength increases?

What happens to photon energy as wavelength increases?

Construct an argument explaining why long wavelength light is unable to free electrons.

Question 9 Work function. Determine the work function for the metal in the simulation.

17.3 Photoelectric Effect continued

Question 10 Predicting the stopping potential.

What is the work function for the metal?

What is the energy of a 450-nm photon?

Predict the stopping potential required when 450-nm light is focused on the emitting plate.

©1999 Addison Wesley Longman, Alan Van Heuvelen and Paul D'Alessandris

Question 1 Momentum conservation. In a classical collision between two identical billiard balls, one initially moving and the other stationary, is the momentum transferred to the stationary ball larger for a *head-on collision* or a *glancing collision?*

Head-on Collision	
Initial	Final

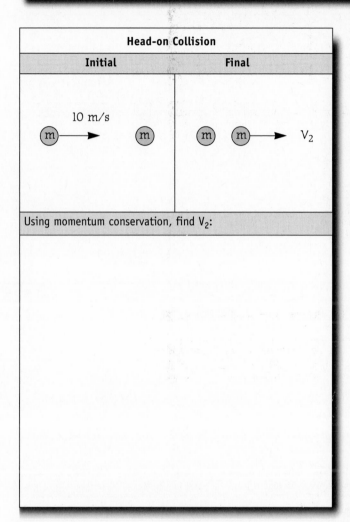

Using momentum conservation, find V_2:

Glancing Collision	
Initial	Final

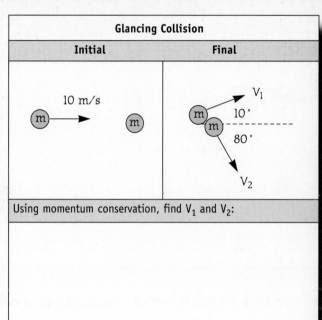

Using momentum conservation, find V_1 and V_2:

Determine the momentum change of the initially stationary ball for both collisions.

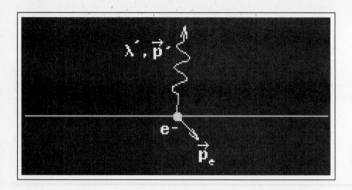

Question 2 Scattered photon wavelength. Predict the wavelength of a 0.035-nm photon after colliding with a stationary electron, if the scattered photon is detected at 83° from its initial direction of travel.

Question 3 Photons scattering from protons. Predict the wavelength of a 0.035-nm photon after colliding with a stationary proton, if the scattered photon is detected at 83° from its initial direction of travel.

Question 4 Resolution of Intensity Peaks. Imagine you have at your disposal a photon detector that can only measure photon wavelength with 2% precision. If you have an incident beam of 0.45-nm x rays, below what angle will your detector not be able to resolve the difference between the scattered and unscattered peaks? Will the scattered photons have wavelengths larger or smaller than 0.45 nm?

What wavelength is 2% longer than 0.45 nm?

Into what angle will photons with the above wavelength be scattered?

Question 5 Resolution vs. incident wavelength. If you use shorter wavelength x rays, will you be able to use your detector at angles smaller than 51°? Does the shift in wavelength ($\lambda'-\lambda$) depend on the initial wavelength, assuming the angle is held constant?

If the wavelength decreases, and the shift remains constant, what happens to the percentage shift of the wavelength?

Will the detector work at less than 51°? Why?

17.4 Compton Scattering continued

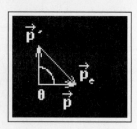

Question 6 Electron momentum. Find the magnitude and direction of the electron's momentum after a collision with a 0.09-nm incident photon, assuming the photon scatters by 90°. (**Hint:** Make use of the momentum diagram.)

Find the photon's initial and final momentum.

Initial: Final:

Find the magnitude of the electrons momentum by trigonometry.

Find the direction of the electron's momentum by trigonometry.

Question 7 Maximum momentum transfer. To provide the greatest momentum transfer to the electron, what wavelength photon should be incident and at what angle should you scatter the photon? (Only use values that the simulation allows.)

Angle
Does the photon transfer more momentum to the electron at small scattering angles or at large ones?
At what allowed angle in the simulation is the maximum momentum transferred to the electron?

Wavelength
Does the incident photon have more momentum at short wavelength or long wavelength?
At what allowed wavelength in the simulation does the photon have the most momentum?
At what allowed wavelength does the photon transfer the most momentum to the electron?

Final answer:

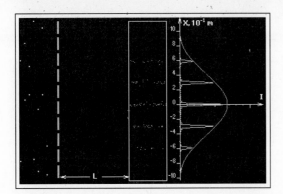

Question 1 Atomic spacing. In order to spread the interference maxima as far apart as possible, should you choose a crystal in which the atoms are spaced by 0.10 nm ($d = 1.0 \times 10^{-10}$m) or 0.15 nm ($d = 1.5 \times 10^{-10}$m)? Explain.

Question 2 Electron's velocity. In order to spread the interference maxima as far apart as possible, should you fire slow electrons or fast electrons? Explain.

Question 3 Minimum resolution. At what speed and atomic spacing will the resulting interference pattern be hardest to resolve, i.e., will have the smallest spacing between maxima? Explain.

Question 4 Analyzing the interference pattern. Set the electron velocity and atomic spacing to generate the largest separation between maxima. Measure the distance between maxima. From this measurement, and the interference relationship used for light, calculate the wavelength of the incident electrons. What velocity and atomic spacing generate the largest separation between maxima?

Velocity:

Spacing:

Measure the distance between maxima:

Calculate the angel between maxima:

Using the diffraction grating equation, calculate the electron's wavelength.

Question 5 The de Broglie relation. What does the de Broglie relation predict for the wavelength of electrons traveling at 1.5×10^7 m/s? Are the effects of special relativity important?

What is the momentum of the electron?

Using the de Broglie relation, calculate the electron's wavelength.

Does the result of the de Broglie relation agree with the result calculated through analyzing the interference pattern?

©1999 Addison Wesley Longman, Alan Van Heuvelen and Paul D'Alessandris

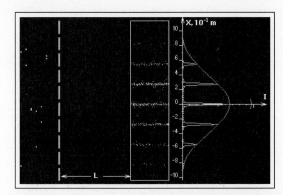

Question 6 Detector placement. If 2.00×10^7 m/s electrons are fired at a crystal lattice with interatomic spacing of 0.123 nm, how far from the central maximum should a detector be placed to record the intensity in the m = 2 peak? Are the effects of special relativity important?

Calculate the electron's momentum.

From the de Broglie relation, calculate the electron's wavelength.

From the diffraction grating equation, calculate the angle of the m = 2 peak.

Calculate the distance of the m = 2 peak from the central maximum.

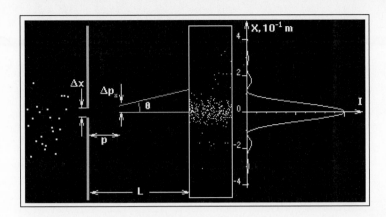

Question 1 The de Broglie relation. What does the de Broglie relation predict for the wavelength of an electron traveling at 2.0×10^7 m/s? What is the electrons momentum?

What is the electron's wavelength?

Question 2 Diffraction minimum. Using the wavelength calculated above, determine the location of the m = 1 diffraction minimum when the electrons pass through an aperture of width 0.30 nm. What is the relationship for determining the location of diffraction minima?

Determine the location of the m = 1 minimum.

Question 3 Uncertainty in x-momentum. When passing through the aperture, the electron's position is specified to be in a region of space of width 0.30 nm. Given the uncertainty in position, what is the minimum uncertainty in x-momentum possible for the electron?

Question 4 Time-of-flight. Now long will it take an electron moving at 2.0×10^7 m/s to reach the screen?

Question 5 X-velocity. Given the x-momentum of our hypothetical electron, what is its x-velocity?

Question 6 X-displacement. Given the time-of-flight and the x-velocity, what is the x-displacement of the electron when it hits the screen? How does this compare to the spread of landing positions predicted by the diffraction equation?

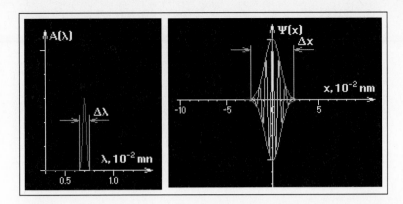

Question 1 Changing the spread of wavelengths. What will happen to the spatial extent of the resultant waveform when a smaller spread of wavelengths are superimposed?

From the de Broglie relation, what effect does a smaller spread in wavelength have on the "spread" in momentum?	From the Heisenberg uncertainty relation, what effect does this have on the spatial extent of the wave form?

What if a larger spread of wavelengths are superimposed?

Question 2 Changing the central wavelength. Will the spatial extent of the resultant wave form change when the spread of wavelengths superimposed are centered on a longer wavelength? Explain.

Question 3 Building a wave packet — I. What is the spatial extent of a wave packet built from plane waves centered around $\lambda = 1.0 \times 10^{-2}$ nm with $\Delta \lambda = 0.2 \times 10^{-2}$ nm?

Question 4 Building a wave packet — II. What spread of wavelengths is necessary to represent an electron with momentum approximately 8.3×10^{-23} kg m/s and spatial extent 4.0×10^{-2} nm?

From the de Broglie relation, what is the approximate wavelength of the electron?

Thus, what spread of wavelengths is necessary to "build" the electron's wave packet?

18

ATOMIC PHYSICS

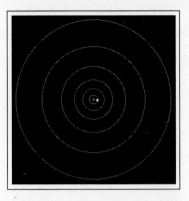

Question 1 Absorption. Does it take more energy for the electron to jump from the ground state to the second orbit or from the ground state to the third orbit? Given your answer, which transition requires a shorter-wavelength photon?

Question 2 Emission. Excite the electron from the ground state to the fifth orbit by absorbing a photon. How will the wavelength of the emitted photon, as the electron returns to the ground state, compare with the wavelength of the absorbed photon, which originally excited the electron into the fifth orbit?

Question 3 Longest wavelength. Which electron transition will emit the longest-wavelength photon?

Question 4 Predicting wavelengths. In order to jump from the second energy level to the fifth energy level, how much energy must the electron absorb? Assuming this energy is transferred to the electron through photon absorption, what wavelength photon must be absorbed? Find the energies of the relevant levels.

Second level:

Fifth level:

Energy difference:

Wavelength of photon:

Question 5 The green line. The transition from the second to the fifth energy level required a photon of wavelength 434 nm to be absorbed, which is blue. Adjacent to this blue line in the spectrum of hydrogen is a green line. This line is also due to a transition involving the second level. What other level is involved in the green line transition, the fourth level or the sixth level? Does the green line represent a longer-wavelength or shorterwavelength photon?

Does the green line represent a higher-energy or lower-energy photon?

Is a higher or lower energy level than the fifth level involved in the transition?

Question 6 Predicting transitions. One day while reflecting on the spectrum of hydrogen, you turn your attention to the infrared line at 1875 nm. What electron transition produces this line? What energy photon has a wavelength of 1875 nm?

What two energy levels differ by an amount equal to the energy calculated above?

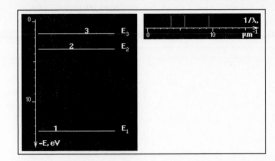

Question 1 Increasing E₂. If the energy of level 2 is increased, how many of the three spectral lines will shift? Which ones? In which direction(s) will each line shift?

λ_{31}:

λ_{32}:

λ_{21}:

Question 2 Decreasing E₃. If the energy of level 3 is decreased, how many of the three spectral lines will shift? Which ones? In which direction(s) will each line shift?

λ_{31}:

λ_{32}:

λ_{21}:

Question 3 Verifying the wavelengths. Given that E_1 = 13.6eV, calculate the three wavelengths of light emitted by the atom you have constructed. Do your calculations agree with the simulation?

$E_1 = -13.6 \ eV$

$E_2 = $ _____

$E_3 = $ _____

$\lambda_{31} = $ _____

$\lambda_{32} = $ _____

$\lambda_{21} = $ _____

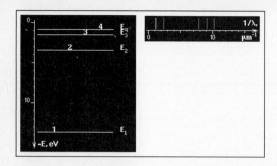

Question 4 Four-level system. How many different wavelengths of light will be emitted by an atom with four distinct electron energy levels? Explain.

Question 5 Increasing E_2 in a four-level system. If the energy of level 2 is increased in a four-level system, which of the *six* spectral lines will shift?

λ_{41}:

λ_{42}:

λ_{43}:

λ_{31}:

λ_{32}:

λ_{21}:

Question 6 Decreasing E_4. If the energy of level 4 is decreased, which of the six spectral lines will shift? In which direction(s) will each line shift?

λ_{41}:

λ_{42}:

λ_{43}:

λ_{31}:

λ_{32}:

λ_{21}:

Question 7 Verifying the wavelengths. Given that $E_1 = -13.6$ eV, calculate the six wavelengths of light emitted by the atom you have constructed. Do your calculations agree with the simulation?

$E_1 = -13.6\ eV$

$E_2 = $ _____

$E_3 = $ _____

$E_4 = $ _____

$\lambda_{41} = $ _____

$\lambda_{42} = $ _____

$\lambda_{43} = $ _____

$\lambda_{31} = $ _____

$\lambda_{32} = $ _____

$\lambda_{21} = $ _____

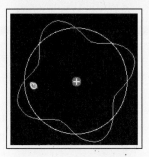

Question 1 Electron phase. Observe the relative phase of the electron as it starts its second circle around the nucleus. Do you think the second pass of the electron around the nucleus will be in phase or out of phase with the first pass? Explain.

Question 2 Standing waves. Does the electron wave interfere constructively, creating a stable standing wave, or destructively, creating a disturbance with an amptitude of zero? Explain.

Question 3 Finding a standing wave. Vary the radius of the electron's path until you find a radius at which the electron will interfere constructively, i.e., form a standing wave. Find as many standing waves as possible. What are the radii of the standing waves? List all stable radii.

Question 4 Allowed wavelengths. Derive a relationship for the allowed wavelengths of electron waves in terms of the radius of the orbit.

Express the relationship between distance traveled and wavelength when the electron arrives at each point along its path in phase with its previous passing.

Express the distance traveled by the electron in one path around the nucleus in terms of the radius of the orbit.

Relate allowed wavelength to radius.

Question 5 Angular momentum. Using the result derived in Question 4, and the de Broglie relation between wavelength and linear momentum, show that Bohr's hypothesis (that the angular momentum of the electron must be a multiple of $h/2\pi$) is simply the result of the electron's wavelike nature.

Express the result of Question 4.

Express the de Broglie relation between wavelength and linear momentum.

Combine these two results and solve for angular momentum $L = mvr$.

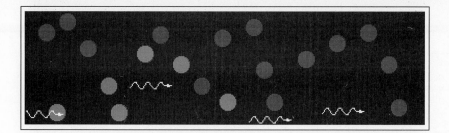

Question 1 Absorption. At any given time, the number of photons inputted into the cavity must be equal to the number that have passed through the cavity without exciting an atom plus the number still in the cavity plus the number of excited atoms. Verify this conservation law by stopping the simulation and counting photons.

$$N_{in} \stackrel{?}{=} N_{art} + N_{remaining} + N_{excited}$$

$$\underline{\quad\quad} = \underline{\quad\quad} + \underline{\quad\quad\quad\quad} + \underline{\quad\quad\quad}$$

Question 2 Direction of spontaneous emission. During spontaneous emission, does there appear to be a preferred direction in which the photons are emitted? Explain.

Question 3 Lifetime of excited state. Does there appear to be a constant amount of time in which an atom remains in its excited state? Explain.

Question 4 Stimulated emission. Carefully describe what happens when a photon interacts with an excited atom. Pay careful attention to the phase and direction of the subsequent photons. (Can you see why this is called stimulated emission?)

Phase:

Direction:

Question 5 Pumping. Approximately what pumping level is required to achieve a population inversion? Remember, a population inversion is when the number of atoms in the excited state is at least as great as the number of atoms in the ground state.

Question 6 Photon emission. Although most photons are emitted toward the right in the simulation, occasionally one is emitted in another direction. Are the photons emitted at odd directions the result of *stimulated* or *spontaneous* emission? Explain.

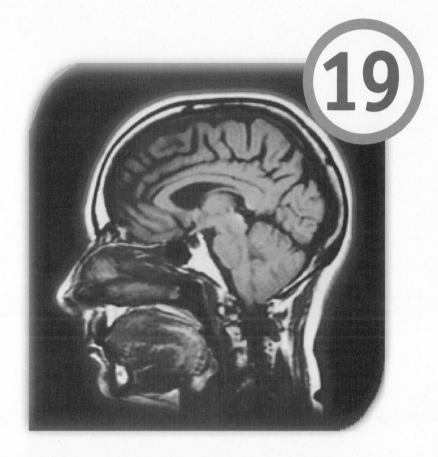

19

NUCLEAR PHYSICS

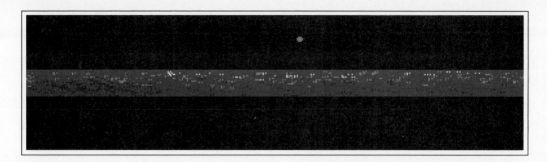

Question 1 Enough information? Does the result of one scattering event provide enough information to identify the radius of the atoms? Explain.

Question 2 Enough information now? Does the result of ten scattering events provide enough information to identify the radius of the atoms? Explain.

Question 3 Constructing a relationship. Construct a relationship between the ratio $N_{unscattered}/N$ and the effective atomic radius. Make sure your relationship agrees with common sense in the appropriate limits.

As the effective atomic radius approaches zero, does your result approach 1.0?

As the effective atomic radius approaches 10 Å, does your result approach zero?

Question 4 R_{atom} =? Rewrite your relationship in the form of "$R_{atom} = $ _____."

Question 5 Precision. Using your scattering data and the relationship derived in Question 4, calculate a tentative value for the atomic radius.

$N_{unscattered} =$

$N =$

R_{atom} (prediction) $=$

If the number of unscattered projectiles had been one larger, what value would the relationship predict for the atomic radius?

In light of these two answers, is your data sufficient to determine the atomic radius to the requested precision of 0.5 Å?

Question 6 Predicting the radius. Using your scattering data and the relationship derived above, predict the atomic radius.

$N_{unscattered} =$

$N =$

$R_{atom} =$

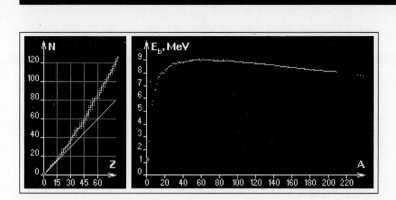

Question 1 **¹H + neutron?** If you add a neutron to a hydrogen nucleus, will the resulting nucleus be stable? What is the name of this nucleus?

Question 2 **¹H + proton?** Create a simple hydrogen nucleus again. If you add a proton, will the resulting nucleus be stable?

Explain why a neutron is "safe" to add while a proton is not.

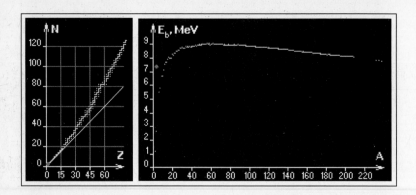

Question 3 ^{3}He? In addition to 2 protons, include a neutron in the nucleus. Is the resulting nucleus stable?

Question 4 ^{4}He? If you add another neutron to the previous nucleus, giving rise to a nucleus with 2 protons and 2 neutrons, will the resulting nucleus be stable? Will it be more or less stable than the previous nucleus?

Question 5 ^{4}He + neutron? Based on what you have learned so far, if you add another neutron to ^{4}He will the resulting nucleus be stable? If so, will it be more or less stable than ^{4}He?

Question 6 N = Z? The following nuclei are stable: ^{2}H, with 1 proton and 1 neutron; ^{4}He, with 2 protons and 2 neutrons; ^{12}C, with 6 protons and 6 neutrons; ^{14}N with 7 and 7, ^{16}O with 8 and 8, and many other nuclei with an equal number of protons (Z) and neutrons (N). Is this a general rule of nature? Is stability guaranteed if the number of protons equals the number of neutrons (N = Z)? Create the following nuclei. Are they stable?

$^{12}_{6}$C

$^{14}_{7}$N

$^{16}_{8}$O

$^{26}_{13}$Al

$^{52}_{26}$Fe

$^{164}_{82}$Pb

Does N = Z guarantee stability?

Question 7 Forming ^{3}He. Determine the amount of energy released if ^{1}H and ^{2}H are fused to form ^{3}He. Determine the binding energy of ^{1}H.

Determine the binding energy of ^{2}H.

Determine the binding energy of ^{3}H.

How much energy is released?

Question 8 Forming ^{4}He. Another reaction taking place in the center of stars is ^{3}He + ^{3}He –> ^{4}He + ^{1}H + ^{1}H. Determine the amount of energy released by this fusion reaction. Determine the binding energy of ^{3}He.

Determine the binding energy of ^{4}He.

Determine the binding energy of ^{1}H.

How much energy is released?

Question 9 Fission. At the other end of the nuclear size spectrum are the extremely large nuclei. Examine the reaction in which uranium fractures into cerium and zirconium.

$$^{235}_{92}\text{U} \longrightarrow \, ^{140}_{58}\text{Ce} \, + \, ^{94}_{40}\text{Zr} \, + \, \text{neuron}$$

Is this reaction allowed by energy considerations? If so, how much energy is released? Determine the binding energy of $^{235}_{92}\text{U}$.

Determine the binding energy of $^{140}_{58}\text{Ce}$.

Determine the binding energy of $^{94}_{40}\text{Zr}$.

How much energy is released?

Initial state: Final state:

^{1}H →   ← ^{1}H ^{1}H ^{1}H

v = 0.03c   v = 0.03c   ← d →

Question 1 Energy bar chart. Construct an energy bar chart for the first half of this collision. Start your analysis when both ^{1}H nuclei are travelling at v = 0.03c and are very far apart and end your analysis when the ^{1}H nuclei instantaneously stop and change directions.

Initial Energy + Work =   Final Energy

K_1 K_2 U_E W K_1 K_2 U_E U_{IN}

Question 2 Energy conservation. Translate your energy bar chart into a mathematical relationship.

Solve this relationship for the *distance of closest approach,* d, the distance between the nuclei when they instantaneously stop.

Question 3 Fusion. In a symmetric collision (both ^{1}H nuclei traveling at the same speed), how large must the speed be to allow fusion to take place? Assume fusion takes place is d $\leq 10^{-15}$m.

Initial state: Final state:

v = 0.04 C   v = 0.04 C

Question 4 Q-value. Calculate the Q-value, the amount of energy released in the fusion reaction.

¹H:

    Mass energy =

    Kinetic energy =

    Total energy =

¹H:

    Mass energy =

    Kinetic energy =

    Total energy =

Initial state:

    Total mass energy =

    Total kinetic energy =

    Total initial energy =

²H:

    Mass energy =

    Kinetic energy =

    Total energy =

e⁺:

    Mass energy =

    Kinetic energy =

    Total energy =

v:

    Mass energy =

    Kinetic energy =

    Total energy =

Final state:

    Total mass energy =

    Total kinetic energy =

    Total final energy =

    Q =

Question 5 Fusion? The ^{1}H + ^{1}H reaction achieves fusion when each ^{1}H nucleus is launched at v = 0.04c. Will the two ^{3}He nuclei also fuse at this speed? Explain.

Question 6 ^{3}He fusion. In a symmetric collision (both ^{3}He nuclei travelling at the same speed), how large must the speed be to allow fusion to take place?

v = 0.032 C   v = 0.032 C

Question 7 Fusion. If the two nuclei are launched at the same speed, 0.032 c, will they undergo fusion?

Find d, the distance of closest approach.

Is d less than 1.0×10^{-15}m? Should fusion occur?

Why does the formula derived in Question 2 not give a correct result?

Initial state:

^{12}C →   ← ^4He

Final state:

^{16}O → γ →

Question 8 A Final Q-value. Calculate the Q-value of the ^{12}C + ^4He reaction.

^{12}C:

Mass energy =

Kinetic energy =

Total energy =

^4He:

Mass energy =

Kinetic energy =

Total energy =

Initial state:

Total mass energy =

Total kinetic energy =

Total initial energy =

^{16}O:

Mass energy =

Kinetic energy =

Total energy =

γ:

Mass energy =

Kinetic energy =

Total energy =

Final state:

Total mass energy =

Total kinetic energy =

Total final energy =

Q =

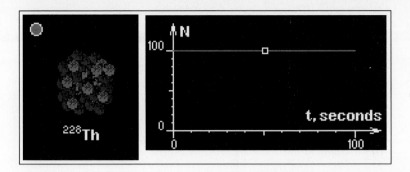

Question 1 Radioactive? *Based only on the data you have collected,* does it appear that ^{228}Th is radioactive? Explain.

Question 2 Decay process. Based on your observations, does ^{228}Th decay via the *alpha, beta-minus,* or *beta-plus* process?

If alpha decay, the resulting nucleus is	If beta-minus decay, the resulting nucleus is	If beta-plus decay, the resulting nucleus is
$^{228}\text{Th} \xrightarrow{\alpha}$ _____	$^{228}\text{Th} \xrightarrow{\beta^-}$ _____	$^{228}\text{Th} \xrightarrow{\beta^+}$ _____

Question 3 Half-life.

Over what time scale does ^{228}Th decay?	Estimate the half-life of ^{228}Th.

Question 4 ^{224}**Ra.** Determine the decay process and estimate the half-life of ^{224}Ra.

If alpha decay, the resulting nucleus is

^{224}Ra $\xrightarrow{\alpha}$ _____

If beta-minus decay, the resulting nucleus is

^{224}Ra $\xrightarrow{\beta^-}$ _____

If beta-plus decay, the resulting nucleus is

^{224}Ra $\xrightarrow{\beta^+}$ _____

Over what time scale does ^{224}Ra decay?

Estimate the half-life of ^{224}Ra.

Question 5 ^{14}**C.** Determine the decay process and estimate the half-life of ^{14}C.

If alpha decay, the resulting nucleus is

^{14}C $\xrightarrow{\alpha}$ _____

If beta-minus decay, the resulting nucleus is

^{14}C $\xrightarrow{\beta^-}$ _____

If beta-plus decay, the resulting nucleus is

^{14}C $\xrightarrow{\beta^+}$ _____

Over what time scale does ^{14}C decay?

Estimate the half-life of ^{14}C.

Question 6 ^{13}**N.** Determine the decay process and estimate the half-life of ^{13}N.

If alpha decay, the resulting nucleus is

^{13}N $\xrightarrow{\alpha}$ _____

If beta-minus decay, the resulting nucleus is

^{13}N $\xrightarrow{\beta^-}$ _____

If beta-plus decay, the resulting nucleus is

^{13}N $\xrightarrow{\beta^+}$ _____

Over what time scale does ^{13}N decay?

Estimate the half-life of ^{13}N.

Initial state: Final state:

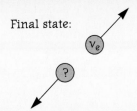

p^+ →   ← e^-

Question 1 Initial state. What are the initial electric charge, baryon number, electro-lepton number, muon-lepton number, and tau-lepton number for this interaction?

Proton <____, ____, ____, ____, ____>

Charge: _____

baryon number: _____

e–lepton number: _____

μ–lepton number: _____

τ–lepton number: _____

Electron <____, ____, ____, ____, ____>

Charge: _____

baryon number: _____

e–lepton number: _____

μ–lepton number: _____

τ–lepton number: _____

Initial state: <____, ____, ____, ____, ____>

Question 2 Unknown particle. Determine the "numbers" on the unknown outgoing particle. From this information, determine the identity of the unknown particle.

Electron neutrino <____, ____, ____, ____, ____>

Charge: _____

baryon number: _____

e–lepton number: _____

μ–lepton number: _____

τ–lepton number: _____

Unknown particle <____, ____, ____, ____, ____>

Charge: _____

baryon number: _____

e–lepton number: _____

μ–lepton number: _____

τ–lepton number: _____

Unknown particle
identity is _____.

Initial state:

Final state:

Question 3 Unknown particle, once again.

Antiproton <_ _ _ _ _>

Antineutron <_ _ _ _ _>

Antimuon <_ _ _ _ _>

Unknown particle <_ _ _ _ _>

Initial state <_ _ _ _ _>

Unknown particle identity is: _____

Initial state:

Final state:

Question 4 Unknown particle, one last time.

Antiproton <_ _ _ _ _>

Antidelta++ <_ _ _ _ _>

Pion <_ _ _ _ _>

Unknown particle <_ _ _ _ _>

Initial state <_ _ _ _ _>

Unknown particle identity is: _____

QUANTUM MECHANICS

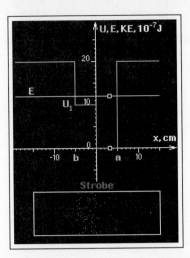

A particle of energy 12×10^{-7} J moves in a region of space in which the potential energy is 10×10^{-7} J between the points −5 cm and 0 cm, zero between the points 0 cm and +5 cm, and 20×10^{-7} J everywhere else.

Question 1 Range of motion. What will be the range of motion of the particle when subject to this potential energy function?

Question 2 Turning points. Clearly state why the particle cannot travel more than 5 cm from the origin.

Question 3 Probability of detection. Assume we measure the position of the particle at several random times. Is there a higher probability of detecting the particle between −5 cm and 0 cm or between 0 cm and +5 cm?

In which region is the particle traveling slower?

In which region does the particle spend more time?

In which region is the particle more likely to be detected?

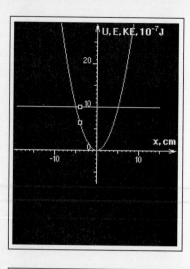

Turn off the strobe display and select the Harmonic potential. Select a particle energy of 10×10^{-7} J and a spring constant of 8×10^{-4} N/m. Run the simulation.

Question 4 Range of motion.

What is the range of motion of the particle?

What will happen to the range of motion of the particle if its energy is doubled?

Question 5 Kinetic energy.

Sketch the graph of the particle's kinetic energy vs. position.

KE

x

Explain why the graph has this shape.

Question 6 Most likely location(s). Assume the position of the particle is measured at several random times.

Where is the particle traveling the slowest?

Where is the particle most likely to be detected?

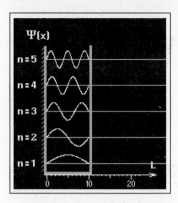

Question 1 Standing waves. From your study of mechanical waves, what is the longest wavelength standing wave on a string of length L?

Question 2 The de Broglie relation. What is the momentum of the longest-wavelength standing wave in a box of length L?

What is the de Broglie relation between momentum and wavelength?

Combine the de Broglie relation and the expression for the longest standing wavelength to relate the momentum to the box length, L.

Question 3 Ground-state energy. Assuming the particle is not traveling at relativistic speeds, determine an expression for the ground-state energy.

What is the relation between particle energy and momentum, in a region of no potential energy?

Combine the above relation and the result from Question 2 to determine the energy of the longest standing wave.

Explain why the longest standing wave has less energy than all other standing waves.

Question 4 Increasing L. If the size of the box is increased, will the ground-state energy increase or decrease?

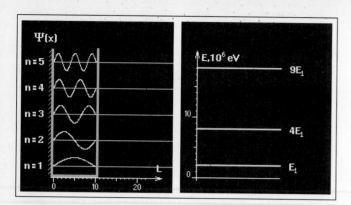

Question 5 The correspondence principle: Large size.
In the limit of a very large box, what will happen to the ground-state energy and the spacing between allowed energy levels? Explain.

Explain why quantum effects are not noticeable in everyday, macroscopic situations.

Question 6 The correspondence principle: Large mass.
In the limit of a very massive particle what will happen to the ground-state energy and the spacing between allowed energy levels? Explain.

Explain why quantum effects are not noticeable in everyday, macroscopic situations.

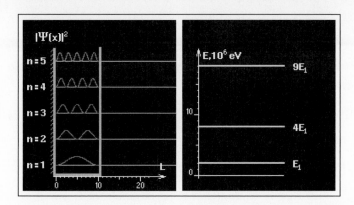

Question 7 Ground-state probability. If a measurement is made of the particle's position while in the ground state, at what position is it most likely to be detected?

Question 8 Probability: Dependence on mass and size. The most likely position to detect the particle, when it is in the ground state, is in the center of the box. Does this observation depend on either the mass of the particle or the size of the box?

Question 9 Probability: Dependence on energy level. The most likely position to detect the particle, when it is in the ground state, is in the center of the box. Does this observation hold true at higher energy levels?

Question 10 The correspondence principle: large n. In the limit of large n, what will happen to the spacing between regions of high and low probability of detection?

Explain why quantum effects are not noticeable in everyday, macroscopic situations.

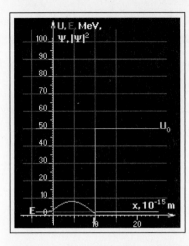

Question 1 Infinite well. If the potential well is infinitely deep, determine the ground-state energy.

Is this also the ground-state energy in the finite well?

Question 2 First excited state. If the potential well is infinitely deep, determine the energy of the first excited state (n = 2).

Is this also the energy of the first excited state in the finite well?

Question 3 "Forbidden" regions. Since the wave function can penetrate into the "forbidden" regions, how will the wavelength of the wave function in the finite well compare to the wavelength of the wave function in the infinite well?

Will the energy of the first excited state in the finite well be greater than or less than the energy of the first excited state in the infinite well? Why?

Question 4 More shallow well. If the depth of the potential well is decreased from 50 MeV to 25 MeV, what will happen to the penetration depth?

What will happen to the wavelength of the wave function?

What will happen to the energy of the n = 3 state?

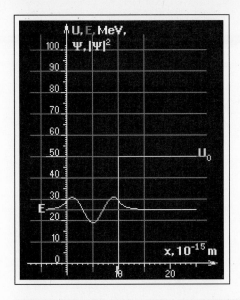

Question 5 Penetration depth (δ). Complete the following table.

m	δ
0.5	
1.0	
2.5	
3.5	
4.5	
(m_p)	($\times 10^{-15}$m)

What happens to the penetration depth as the mass of the particle is increased?

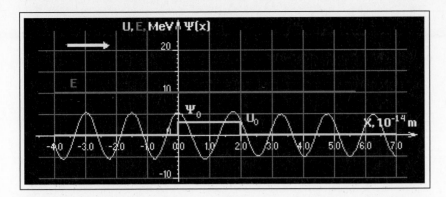

Question 1 Creating a potential barrier. If a potential barrier of a few MeV is created, what will happen to the transmission probability of the particle? Explain.

Question 2 Raising the barrier. What will happen to the transmission probability when the potential equals, and then exceeds, the initial energy of the particle? Will the particle be able to be transmitted through the barrier? Explain.

Question 3 Barrier width. If the width of the barrier is decreased, what effect will this have on the transmission probability? Explain.

20.4 Potential Barriers continued

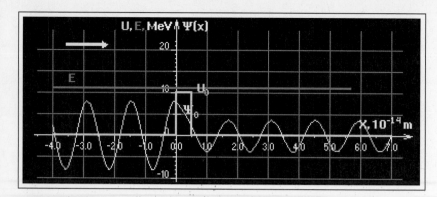

Set the potential height to 11 MeV, the incident energy to 10 MeV, and the barrier width to 0.5×10^{-14}m.

Question 4 Functional effect of barrier width on transmission probability.

Record the transmission probability.	Sketch a graph of T vs. L.

Double the barrier width and record the new transmission probability. Continue to double the width and record the new transmission probabilities.

L	T
0.5	
1.0	
2.0	
4.0	
($\times 10^{-14}$m)	

What is the approximate functional form of the dependence of T on L? Explain.

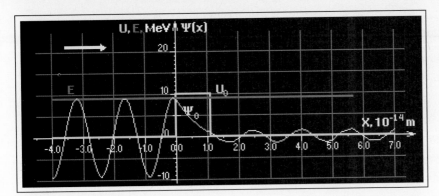

Set the potential height to 9 MeV, the incident energy to 10 MeV, and the barrier width to 1.1×10^{-14} m.

Question 5 Tricky question. If the barrier width is slowly increased to 2.3×10^{-14} m, what will happen to T? Describe what you discover.

Question 6 Doubly tricky question. If you continue to increase the barrier width to 4.7×10^{-14} m, what will happen to T? Describe what you discover.

Question 7 Explanation. From your careful observation of the wave function, formulate an explanation as to why the transmission probability does not always decrease when the barrier gets wider.

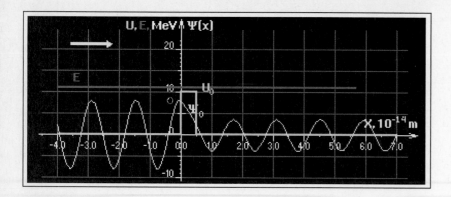

Question 8 Scattering from potential wells. What will happen to the transmission probability if the potential is dropped to a negative value, i.e., a potential well is formed? Can the particle get reflected from a potential well? Explain.

Question 9 No Longer a Tricky Question. Continue to lower the potential energy in the well. Why does the transmission probability have maximas at −4 MeV and −10 MeV? Explain.

Activ**Physics** 2
User Guide

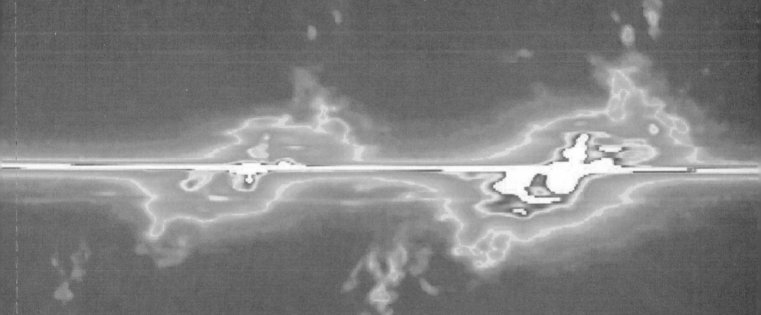

About *ActivPhysics*

ActivPhysics 2 is a collection of guided, interactive activities that present Electricity and Magnetism, Optics and Modern Physics for use in introductory college or university physics courses or high school courses that emphasize problem solving. All the activities are simulation based.

ActivPhysics 2 contains many features that convey concepts and information in a visual, interactive way, unlike most traditional methods.

This CD-ROM can be used alone, but the accompanying *ActivPhysics 2* Workbook is strongly recommended.

Navigating *ActivPhysics*

Main
Brings you back to the main menu of the unit you are in.

ActivPad
Launches *ActivPad,* an interactive notebook you can use to keep notes and create customized links to any place in *ActivPhysics 2* or the World Wide Web.

Help
Outlines navigation features and provides tips on how to integrate the product into your course.

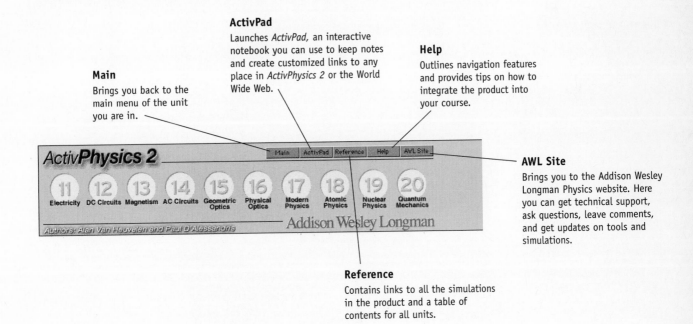

AWL Site
Brings you to the Addison Wesley Longman Physics website. Here you can get technical support, ask questions, leave comments, and get updates on tools and simulations.

Reference
Contains links to all the simulations in the product and a table of contents for all units.

Units and Activities

You can move to a unit by selecting one of the unit names or numbers along the middle of the menu. Once you have made your selection, the titles of the activities for that unit will appear in the frame on the right side (see below). Simply click on an activity to launch it. The activity will appear in the frame on the right, and all simulations will appear on the left.

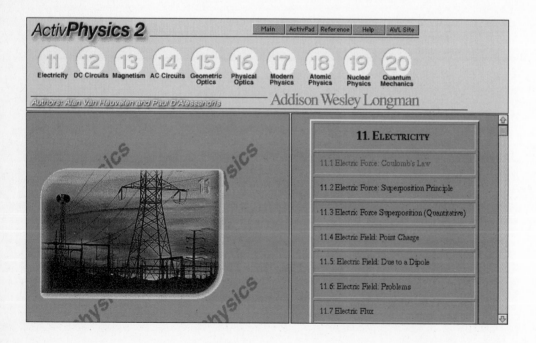

©1999 Addison Wesley Longman, Alan Van Heuvelen and Paul D'Alessandris

Navigating an *ActivPhysics* Activity

The Help button in the main menu brings you to a sample interactive activity that illustrates what each button link in an *ActivPhysics* activity does. This interactive Help activity is a great way to learn how to use the *ActivPhysics* interface. It is recommended that you review this section before working with an *ActivPhysics* activity.

   Objective button links to a description of the goals of the activity.

   Simulation button launches interactive simulation for the activity.

   Advisor button links to information and tips to help you complete the activity.

Some *ActivPhysics* activities contain computer tools to help you solve problems and make predictions:

Force Diagram Tool:
Allows you to create force diagrams on the screen. Note that the tails of the force arrows should attach at the place the flow is exerted on the object of interest.

Bar Chart Predictor:
Allows you to qualitatively predict the initial and final work-energy terms in a work-energy process.

Using the Simulations

 Simulations are launched by clicking on this icon.

The simulations contain sliders and controls that allow you to alter the conditions of the physical system represented by the simulation. Simulations can be accessed through the on-screen activity or through the Reference section.

As you roll the cursor over the simulation screen, information about the simulation, including descriptions of what the various controls do, appears at the bottom of the interface screen.

To get simulation help, click anywhere on the simulation. (In Windows, right mouse click).

Using ActivPad

ActivPad is an on-line, interactive notepad you can use to keep notes and "bookmark" places in *ActivPhysics*. You can also use *ActivPad* to create hyperlinks to sites on the World Wide Web. *ActivePad* is ideal for creating lecture notes or custom homework problems, because you can have instant hyperlinks to *ActivPhysics* tools or other Web sites.

Students Use *ActivePad* to:

keep your notes about the various activities, questions you want to ask an instructor or TA later, etc.;

post your answers to *ActivPhysics* questions and paste in material from *ActivPhysics*, or graphs and charts from other programs;

"bookmark" places in *ActivPhysics* and, if you have a live Web connection, to add links to physics sites or other related material on the Web. You can also link to your course home page if one has been posted.

Instructors Use *ActivPad* to:

Organize your lecture notes and class demonstrations. Type your notes into *ActivPad* and/or use the active linking feature to create links to the simulations or other portions of *ActivPhysics* that you want to show in class.

Create new questions or problems to accompany the existing *ActivPhysics* simulations. You can save your *ActivPad* assignment to a server where your students can access it.

Save links to physics resources on the Web, including your course home page.

To Set Up *ActivPad* on Windows

In Netscape 4:

1. With *ActivPhysics* 2 open, from the **Edit** menu select **Preferences.**
   A Preferences window opens.

2. Under **Category** at the left side of the Preferences dialog box, select **Navigator** by clicking on the plus symbol. Then, click on the **Applications** sub-category to open the Applications dialog area on the right side of the Preferences window.

3. On the right side of the dialog box, click the button labeled **New Type...**

4. In the dialog box enter:
   Description of Type: **ActivPad**
   File Extension: **led**
   MIME Type: **application**
   Application to use: Click the **Browse** button and locate *ActivPad* 1.0 by browsing your hard drive for the application.

5. Click **OK** to close the dialog box and return to the Preferences window.

6. Close the Preferences dialog box by clicking OK.

In Internet Explorer 4:
Internet Explorer 4.01 on the Windows platform cannot be configured to automatically launch *ActivPad*. If you plan to use this application while running *ActivPhysics* 2, please locate and launch *ActivPad* from your hard drive. For more information, please click the "AWL Site" button on your toolbar in *ActivPhysics*.

To Set Up ActivPad on Macintosh

In Netscape 4:

1. With *ActivPhysics* open, from the **Options** menu select **General Preferences.**
   A Preferences window opens.

2. Select the **Helpers** tab and click the **[New]** button.

3. In the dialog box type:
   Description: ActivPad
   MIME Type: **application/led**
   Suffixes: **led**
   Select **Application** in the **Handled By** box

4. Select *ActivPad* as the application by clicking the Browse button to locate the *ActivPad* application on your hard drive and then clicking **Open.**

5. Click the File Type pop-down menu and select **LDDC.**

6. Click OK to close the Helpers dialog box.

7. Click OK to close the Preferences window.

In Internet Explorer 4:

1. With *ActivPhysics* open, from the **Edit** menu select **Preferences.**
    A Preferences window opens.

2. Locate the **Receiving Files** category in the list on the left side of the window and click the arrow to open the sub-menu, if it is not already open. Under Receiving Files, select **File Helpers.**

3. In the File Helper Settings dialog area on the right, click the **Add...** button. A new dialog box will open titled **Edit File Helper.**

4. In the area of the dialog box labeled **Representation** type:
    Description: ActivPad
    Extension: **led**
    MIME Type: **application/led**

5. In the area of the dialog box labeled **File Type,** click the **Browse...** Locate the *ActivPad* application on your hard drive and then click **Open.** The File type field should automatically be filled in with the type LDDC and the File creator field should read Apad.

6. In the bottom area of the dialog box labeled **Handling,** open the pull-down menu and select View With Application. The application name and icon should then appear below the menu. If it does not, click the Browse... button in the Handling area and repeat the procedure in item 5 above.

7. Click OK to close the Edit File Helpers dialog box.

8. Click OK to close the Preferences window.

Managing Your *ActivPad* Documents

ActivPad documents are like any other documents that you save on your computer. You can save them all in one folder, or in various locations. You decide where an *ActivPad* document will be stored when you name and save it.

If you work with *ActivPhysics* and *ActivPad* on more than one computer, be sure to see "Sharing *ActivPad* Documents."

Taking Notes in *ActivPad*

Note taking using *ActivPad* is straightforward. The program behaves much like a regular word processor.

1. In the **Main** menu bar, launch *ActivPad* by clicking on the *ActivPad* button.

ActivPad does not launch from within Internet Explorer. If *ActivPad* is not launched from the Main Menu, locate the *ActivPad* application on your hard drive and launch it from there.

Cutting and Pasting in *ActivPad*

You can cut and paste material from other documents into *ActivPad*. This includes material from *ActivPhysics*, word processing documents, clip art, graphics, charts and tables from spreadsheets, etc. Simply open the document you want to paste from, copy the material, then open your *ActivPad* document and paste the material in.

ActivPad is not a full math processor, so check the pasted material for accuracy when cutting and pasting formulas or other math text.

Depending on your computer's memory allocations, you may not be able to have many documents open at once. You may have to open and close documents while copying and pasting.

Windows users using Netscape cannot paste images into *ActivPad* directly from *ActivPhysics*; those using Internet Explorer can do so.

Creating a Link in *ActivPad*

Make sure that both *ActivPad* and *ActivPhysics* (or the Web document you want to link to) are open. It is easiest to create the link if you can see the *ActivPad* document you are working with and the document you want to link to at the same time. To do this you can reduce the size of the windows of each application in such a way that they are side by side or one on top of the other.

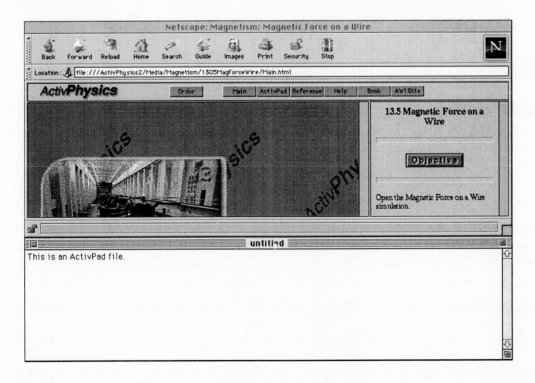

Grab Current

Grab Current is a method to "bookmark" useful resources in *ActivPhysics* or on the Web.

1. With *ActivPad* open, go to the spot in *ActivPhysics* or on the Web to which you want to create an *ActivPad* link.

2. From the ActivLinks pull-down menu, choose **Grab Current URL.** The link will appear in your *ActivPad* document.

Drag and Drop

Drag and Drop is helpful when you want to grab many links from one page and put them into a single *ActivPad* document.

1. With *ActivPad* open, select the place in *ActivePhysics* or on the Web that contains the link you want to add to *ActivPad*. *Do not activate the link.*

2. Select the link and drag it over to the open *ActivPad* document and release the mouse. The link will appear in your *ActivPad* document.

3. Double-click on the new link to activate it.

You need to have either *ActivPhysics* or a Web connection open in order for your links to work (depending on where they point to).

You can delete any links that you have created in *ActivPad* by simply selecting them and hitting the delete key.

ActivPad links will open any framed links full screen instead of in their original framed window.

Customizing Your *ActivPad* Links

When you bring a link into *ActivPad* from *ActivPhysics* or the Web you will get a link label that shows the title of the link, information about the link's point of origin, etc.

You can rename the label or customize it so that the link is represented by an image. You may want to do this to make the link easier to remember, to connect an image to a link for study purposes (e.g., an image of a graph linking to material about the physical phenomena it represents), or to make your *ActivPad* document visually interesting if you are using it for demonstration purposes.

Creating a Link Represented by an Image

1. Paste the image you want to use to represent a link into the *ActivPad* document (See "Cutting and Pasting in *ActivPad*" for help with this.)

2. Select the place in *ActivPhysics* or on the Web that contains the link you want to connect to the image. *Do not activate the link.*

3. Select the link (Right click for Windows; click and hold for Mac) until the pop-up link menu appears. From the menu, select **Copy this Link Location.**

4. Return to the image you pasted in the *ActivPad* document. Select the image.

5. From the **ActivLinks** menu, select **Edit ActivLink.** A new window will open.

6. Highlight all of the text in the URL line of this window, and then select **Paste.** This replaces the old link information with the new information you copied from your desired link. If there is no text in this line simply select **Paste.**

7. You have now created an image link in *ActivPad*, which works like every other *ActivPad* link. Remember that the document you are linking to must be open, whether it is *ActivPhysics* or a web browser.

Changing a Link's Label

1. Select the link in *ActivPad*.

2. Go to the menu bar at the top of the screen and open the **ActivLinks** menu. From that menu, select **Edit ActivLink.** A new window will open.

3. Replace any text in the title line of this window with whatever text you want to change it to.

4. Your link still goes to the same location it originally did, but now has the title of your choice. Remember that the document you are linking to must be open, whether it is *ActivPhysics* or a web browser.

Sharing *ActivPad* Documents

Standardizing *ActivPad* Links

If you intend to use *ActivPad* documents on a different computer or to share *ActivPad* documents with other *ActivPhysics* users, you should use *ActivPad's* standardizing feature.

When an ActivLink is created, *ActivPad* records a URL, or digital address, so that the program will later be able to retrieve that same location on the CD-ROM or Web. Because different computers are configured differently, this address varies from computer to computer. While one computer may have the CD-ROM drive designated as drive D, another may call it drive G, and yet another (a Macintosh, for example) does not use a letter to denote its CD-ROM drive. Thus the same *ActivPhysics* home page may have several different addresses, depending on the computer it is being viewed on:

> Your computer: file///D|/ActivPhysics2.html
> PC 2: file: ///E|/ActivPhysics2.html;
> Web server: http://www.phys.university.edu/ActivPhysics2.html
> Macintosh: file://ActivPhysics_2_CD/ActivPhysics2.html

ActivPad's standardizing feature will automatically replace the first part of these URL's with a variable name like "$ActivPhysics2$." Thus, the URLs from above will be recorded in *ActivPad* as follows:

> Your computer: $ActivPhysics2$ActivPhysics2.html
> PC 2: $ActivPhysics2$ActivPhysics2.html
> Web server: $ActivPhysics2$ActivPhysics2.html
> Macintosh: $ActivPhysics2$ActivPhysics2.html

Now all standardized links will work properly when shared among the above users.

Make sure that everyone is using the same variable name.

You do not need to standardize links to the World Wide Web.

Standardizing ActivPad Hyperlinks

You should standardize *ActivPad* hyperlinks if you intend to do any of the following:

Share *ActivPad* documents with other users.

Distribute *ActivPad* documents as class notes or custom homework sets.

Use *ActivPad* documents on multiple computers.

You must perform the following steps with the *ActivPhysics* home page open.

To Standardize *ActivPad* Documents

1. Launch *ActivPad* by clicking the **ActivPad** button on the Main Screen of *ActivPhysics*.

2. From the **ActivLinks** menu choose **ActivLink Preferences.** A window appears listing *ActivPhysics* in a **Variable** box.

3. Click on the **Grab** button to the right of the **ActivPhysics Variable box.**

4. Click the check box under the **Standardize New URL's** heading, if a check is not already there.

5. To test: From the **ActivLinks** menu, select **Grab Current Location.** The following hyperlink should be inserted into the current *ActivPad* document:

   URL:$ActivPhysics2$ActivPhyiscs2.html

If the link is not inserted properly, check your **ActivLink Preferences** to make sure that the **Standardized new URL's with this prefix** checkbox is checked.

Getting Help with *ActivPhysics*

If you find that you need help or technical support:

Try reading the section of this User Guide that covers the part of *ActivPhysics* about which you have a question. There is also a README file and a Help file on the *ActivPhysics* CD-ROM. You may find an answer to your question in one of these.

Addison Wesley's physics web site (http://www.awl.com/physics) is also a good source for updates on new products, technical information, frequently asked questions, and user comments.

If your problem is related to installation or defective media, please contact Addison Wesley's Technical Support Division via e-mail at mailto:techsprt@awl.com, or by phone at 800-677-6337. The hotline is staffed from 9 a.m. to 4 p.m., Monday through Friday (Eastern time). If you do not get through immediately, please leave them a message describing your problem. A software technician will return your call. Please note that Technical Support provides installation guidance and defective media replacement only. Questions on program usage should be directed to Addison Wesley Longman Physics via our web site. If you are a student, you can also direct questions on program usage to your professor or lab coordinator.